The Geology of Yellowstone

A Biblical Guide

By Patrick Nurre

The Geology of Yellowstone
A Biblical Guide

By Patrick Nurre

The Geology of Yellowstone – A Biblical Guide
Published by Northwest Treasures
Bothell, Washington
425-488-6848
NorthwestRockAndFossil.com
northwestexpedition@msn.com

Cover photo: Grand Prismatic Spring. Photo by Vicki Nurre.

The Geology of Yellowstone
A Biblical Guide

Contents

Preface 5
Chapter One – A Short History of Yellowstone National Park 6
Chapter Two – An Introduction to the Geology of Yellowstone 16
Chapter Three – Interpreting Yellowstone – The Biblical Geology of Yellowstone 27
Chapter Four – The Yellowstone Supervolcano 58
Chapter Five – The Thermal Features of Yellowstone 66
Chapter Six – The Microbes of Yellowstone 75
Chapter Seven – The Rocks and Minerals of Yellowstone 86
Chapter Eight – The Ice Age in Yellowstone 122
Chapter Nine – A Brief Road Guide to Yellowstone 136
- Lewis and Clark Caverns 137
- Hebgen Lake/ Earthquake Lake 139
- The Grand Loop 144
- The West Entrance 144
- North to Norris Geyser Basin 149
- Norris Geyser Basin 152
- Roaring Mountain 158
- Obsidian Cliff 158
- Sheepeater Cliff 160
- Bunsen Peak 161
- Mammoth Hot Springs 164
- Toward the North Entrance 166
- East to Tower/Roosevelt 171
- East to the Northeast Entrance 171
- Specimen Ridge 171
- The Scenic Beartooth Highway 176
- The Chief Joseph Highway 180
- South at Tower/Roosevelt 182

Canyon Village/Grand Canyon of the Yellowstone 184
South toward Lake Village/Fishing Bridge 185
Southwest toward West Thumb/Grant Village 188
West Thumb Geyser Basin 190
South toward the South Entrance 193
West toward Old Faithful 194
Upper Geyser Basin 195
Old Faithful 196
North toward Midway Geyser Basin 204
Grand Prismatic Spring 204
North toward Lower Geyser Basin 205
Fountain Paint Pots 206
North to Madison Junction 209
Picture Credits 210

Preface

The various interpretations and conflicting dates for the origin of many of the features of Yellowstone present a confusing picture of the secular geological perspective of Yellowstone, to say the least. There have been hundreds of books and field guides written on Yellowstone. My guide is an attempt to present the geology of Yellowstone from a Biblical perspective and to help you enjoy this amazing wonderland of geological phenomena within a Biblical framework. It is certainly not exhaustive. The guide brings out the geology that has especially interested me. The geology of Yellowstone covered in this guide is also some of the most perplexing to modern geological interpretations, though most people would not know this.

I have led numerous field trips to Yellowstone. The response to learning the Biblical geology of Yellowstone has always been an eye-opening experience. People just do not realize that there are two ways to interpret the geology of Yellowstone, and both depend on a prior commitment to some idea about earth history. Which one is right? Because I believe that God has revealed a history of the earth in the first 11 chapters of Genesis, then it follows that I believe that the Biblical view of a recent creation and global flood is the right one. I have therefore written this guide with that prior commitment.

Patrick Nurre
Bothell, Washington
2014

Chapter One
A Short History of Yellowstone National Park

In 1872 President Ulysses S. Grant signed into law the establishment of the world's first national park – Yellowstone National Park. It has served as a worldwide model ever since for the setting aside of lands for historical preservation and enjoyment of the people.

But the story begins long before this. The Corps of Discovery led by Lewis and Clark in 1804 through 1806 was commissioned by President Thomas Jefferson to explore and document as much information as they could about the great Louisiana Purchase, recently acquired from France in 1803. The purchase included lands that now incorporate Yellowstone National Park. Although Lewis and Clark explored the greater Yellowstone region and much of the Yellowstone River, they never entered the region now known as Yellowstone Park. On the return trip of the Lewis and Clark Expedition, John Colter, one of the Corps of Discovery scouts separated from the Corps and spent the winter of 1807-1808 exploring the now famous national park region. John Colter is generally considered to be the first person of European descent to enter the Yellowstone country.

John Colter

Colter returned to the eastern United States with fantastic stories about waterfalls, *fire and brimstone*, and boiling hot pools. Many people thought he had gone crazy. His mythical Yellowstone was called *Colter's Hell.*

Colter's fantastic Yellowstone would have to wait for future discoveries to bring this wonderful place to the public attention.

There were some who were interested, though, mainly trappers and miners. The next several years saw many of these explorers come and go. They continued to bring fantastic stores of geological wonders from this isolated wilderness.

It was Dr. Ferdinand Hayden, geologist, who organized the first scientific exploration of Yellowstone, who in 1871 petitioned Congress for $40,000 to fund the expedition. What Hayden brought back in the way of maps, samples and surveys prompted Congress to set aside this geological wonderland as a protected park.

From an article by Dr. Hayden, illustrated by Thomas Moran

The Hayden Expedition – The first official expedition to Yellowstone

There were two men that Hayden had invited to be a part of the Yellowstone expedition in 1871. One was the artist Thomas Moran. The other was the photographer William Henry Jackson. It was their paintings and photographs of Yellowstone that probably did more than anything else to help in the creation of Yellowstone as a National Park. Moran's paintings have become a very popular part of telling the story of Yellowstone. And their work has become iconic.

Photographer William Henry Jackson of Civil War fame
and artist Thomas Moran

Thomas Moran's *The Grand Canyon of the Yellowstone*, 1871

Jackson's photograph of Tower Fall, 1871

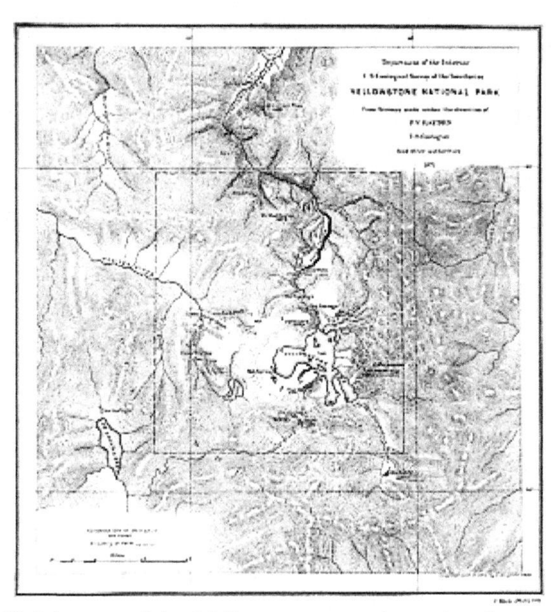

The first official map of the Yellowstone made under the supervision of Hayden in 1871

Just after the Park was commissioned in 1871, the United States Army was given the responsibility for managing the Park and protecting its resources. Its primary job during those early years was spent arresting and preventing poachers who had come to the Yellowstone to hunt buffalo and other game. In 1917, administration of the park was transferred to the National Park Service, which had been created the previous year.

Bison heads confiscated from poachers

Men in post exchange at Ft. Yellowstone

The first military outpost in Yellowstone was located at Norris Geyser Basin. It is still situated there, but now serves as a museum.

After the newly established National Park Service took over administration of Yellowstone, the Union Pacific Railroad began investing in laying of tracks to Gardiner, Montana and West Yellowstone, Montana, and in building a hotel at Mammoth Hot Springs.

Early advertisement for Union Pacific: Stagecoach picking up passengers at the newly built Mammoth Hotel inside the north entrance to Yellowstone

Train station at Gardiner, Montana where passengers were dropped off for their pickups into Yellowstone National Park by stagecoach

Some claim that Henry G. Merry made Yellowstone history when he drove his Winton Motor Carriage into the park in 1902. He is credited with driving the first car into Yellowstone. Although cars were not officially allowed into the Park until 1915, that didn't stop the more adventurous.

Today Yellowstone hosts over 3,000,000 visitors a year; and it is not the most visited National Park! Great Smoky Mountains National Park holds that honor at over 6,000,000 visitors a year! In fact, Yellowstone is number four on the list of most visited National Parks in the United States. And Yellowstone is not even the biggest National Park in the US.

The first four spots in that list are held by National Parks in Alaska with a total of 37,910 square miles of National Park land – roughly nine times the size of Yellowstone! Yellowstone is number eight on the list with over 3,472 square miles of designated protected land or 2,221,766 acres. 96% of the Park is in Wyoming, 3% in Montana and 1% in Idaho. The Park is Larger than Rhode Island and Delaware combined. Approximately 5% of the park is covered by water, 15% is grassland and 80% is forest. Despite its small size in comparison to some of the other National Parks in the US, it is filled with over 10,000 thermal features, 300 geysers (the largest concentration in the world), 290 waterfalls, the largest fresh water lake above 7,000 feet, and has 1,000-3,000 earthquakes per year.

Besides being a wonderland of nature, it is also an amazing school for the study of geological forces brought on by the Genesis Flood and its aftermath. Consider the following:

- ⇨ The Yellowstone caldera, home of the largest volcanic eruptions in earth history, is an active volcano. The most travelled portion of Yellowstone is within one of the calderas that encompasses a length of 45 miles and a width of 35 miles. It is a massive volcano that if it erupted again would take out everything within a 250-mile radius!
- ⇨ It is the hottest and most acidic natural place on earth housing hot springs with a pH of nearly 1.3 – the concentration of battery acid.
- ⇨ The last major eruption of Yellowstone erupted an estimated 240 cubic miles of rock, ash and other material. To put this in focus, Mt. St. Helens erupted just a little more than .25 cubic miles of volcanic material.
- ⇨ Yellowstone is surrounded by the Absaroka Mountains on the north, the east and southeast. The Absarokas are a mountain chain of extinct volcanoes with an estimated 9,200 cubic miles of volcanic material!
- ⇨ The mountains of the Beartooths, over 10,000 feet in elevation, border Yellowstone National Park, just to the north of the Gallatin and Absaroka Mountains. The Beartooths are composed of granite and gneiss rock. Some of the most outstanding glacial remnant features in the world are in the Beartooth Mountains.

As we work our way through The Geology of Yellowstone National Park and the Greater Yellowstone Ecosystem, we will explore these and many other geologic features that make Yellowstone a geologist's dream!

Some explanation is in order here. The area known as Yellowstone National Park has definite borders and those are typically pictured on various maps showing Yellowstone National Park. But Yellowstone is so much more than just these defined National Park borders. The phrase used to describe the area outside Yellowstone National Park that incorporates the same volcanism, Ice Age evidence, rocks and minerals, and flora and fauna is called The Greater Yellowstone Ecosystem.

A 1938 Yellowstone advertisement poster

Chapter Two
An Introduction to the Geology of Yellowstone

For years, we have been told that Yellowstone National Park is the result of catastrophic eruptions of a series of super volcanoes that began some 16 million years ago. Most people today accept this without a second thought. Since scientists say it, it must be right. I mean, after all, they did not just guess at this figure, right? Do scientists know the age of Yellowstone for certain? Scientifically, how did they arrive at their conclusion? In addition, we should be asking how all this relates to the Scriptures. For example, how do the reported ages of Yellowstone square with the Scriptures? If what scientists are saying about the age of Yellowstone is correct, how can we trust the Scriptures, that categorically state that the earth is much younger than what scientists give as the age for Yellowstone?

Staking out a Position

The age for Yellowstone is figured, so the Park literature states, by careful radiometric dating of lavas. While all this sounds scientific and beyond dispute, the whole system of radiometric dating is filled with unverifiable assumptions based on ideas about the presently observed rates of radioactive decay. Because of this, it is necessary to spend some time becoming familiar with this method.

Radiometric Dating

The Bible's chronology makes it clear that the Earth originated about 6,000 years ago by the word of God and that a global catastrophic flood destroyed the Earth about 4,500 years ago. Secular geology claims that the universe is 15 billion years old and the Earth is about 4.6 billion years old. If this is true, then the entire Bible is a lie! Furthermore, secular geologists claim to have proof positive that the Earth is 4.6 billion years old by way of radiometric dating. Let's look at this and see if it is indeed proof of an old Earth.

What is radiometric dating? Geologists claim that it is a type of natural clock. It is based on the fact of radioactivity, that some elements are unstable and therefore decay. They are radioactive. It is a fact that decay can be measured in the present and this decay rate seems to be constant.

So, according to geologists, all one has to do is to keep track of this present decay and then calculate how long it would take for a radioactive element to decay completely.

An unstable element is one that has too many neutrons in its nucleus. Look at the illustration of the Carbon atom below.

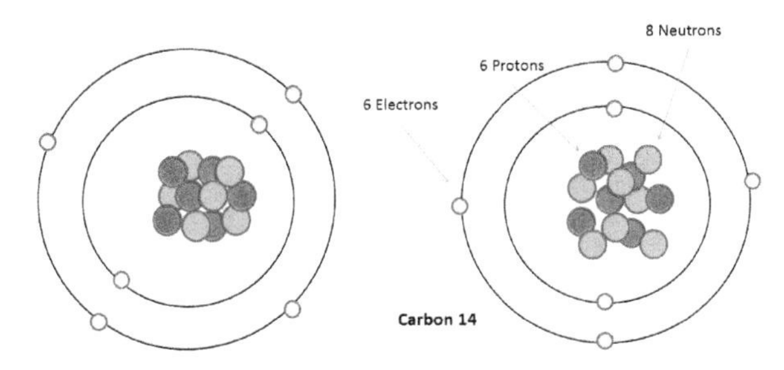

The normal Carbon12 atom (6 protons, 6 neutrons) on the left, and the abnormal (radioactive) Carbon14 atom (6 protons and 8 neutrons): the number of electrons equals the number of protons.

A radioactive element is called an isotope (meaning *same place*) because it occupies the same place as the normal atom on the Periodic Table, but has a different nucleus (too many neutrons, too much energy, or too many protons).

The Periodic Table

The Periodic Table was developed by a Russian named Dmitri Mendeleev in 1869. He developed his table to illustrate periodic trends in the properties of the then-known elements. Mendeleev also predicted some properties of then-unknown elements that would be expected to fill gaps in this table. Most of his predictions were proved correct when the elements in question were subsequently discovered.

The Table is organized based on atomic numbers, electron configurations, and recurring chemical properties. Elements are presented in order of increasing atomic number (number of protons). There are 18 columns. The columns are called groups. The groups are based on atomic structure of the elements; all elements in a group have similar atomic structure.

Periodic Table of the Elements

1	2	3	4	5	6	7	8	9	10	11	12	13	14	15	16	17	18
1 H 1.01																	2 4.00
3 Li 6.94	4 Be 9.01											5 B 10.81	6 C 12.01	7 N 14.01	8 O 16.00	9 F 19.00	10 20.18
11 Na 22.99	12 Mg 24.30											13 Al 26.98	14 Si 28.09	15 P 30.97	16 S 32.07	17 Cl 35.45	18 39.95
19 K 30.10	20 Ca 40.08	21 Sc 44.96	22 Ti 47.88	23 V 50.94	24 Cr 52.00	25 Mn 54.94	26 Fe 55.85	27 Co 58.93	28 Ni 58.69	29 Cu 63.55	30 Zn 65.39	31 Ga 69.72	32 Ge 72.61	33 As 74.92	34 Se 78.96	35 Br 79.90	36 83.80
37 Rb 85.47	38 Sr 87.62	39 Y 88.91	40 Zr 91.22	41 Nb 92.91	42 Mo 95.94	43 Tc (97.91)	44 Ru 101.07	45 Rh 102.91	46 Pd 106.42	47 Ag 107.87	48 Cd 112.41	49 In 114.82	50 Sn 118.71	51 Sb 121.75	52 Te 127.60	53 I 126.90	54 131.29
55 Cs 132.91	56 Ba 137.33	57 La 138.91	72 Hf 178.49	73 Ta 180.95	74 W 183.85	75 Re 186.21	76 Os 190.23	77 Ir 192.22	78 Pt 195.08	79 Au 196.97	80 Hg 200.59	81 Tl 204.38	82 Pb 207.2	83 Bi 208.98	84 Po (208.98)	85 At (209.99)	86 (222.02)
87 Fr (223.02)	88 Ra (226.03)	89 Ac (227.03)	104 Rf (261.11)	105 Ha (262.11)	106 Sg (263.12)												

58 Ce 140.12	59 Pr 140.91	60 Nd 144.24	Pm (144.91)	62 Sm 150.36	63 Eu 151.97	64 Gd 157.25	65 Tb 158.93	66 Dy 162.50	67 Ho 164.93	68 Er 167.26	69 Tm 168.93	70 Yb 173.04	71 Lu 174.97
90 Th 232.04	91 Pa 231.04	92 U 238.03	Np (237.05)	Pu (244.06)	Am (243.06)	Cm (247.07)	Bk (247.07)	Cf (251.08)	Es (252.08)	Fm (257.10)	Md (258.10)	No (259.10)	Lr (262.11)

Above is a representative Periodic Table.

What do the various symbols mean? Let's take a look at the element, uranium.

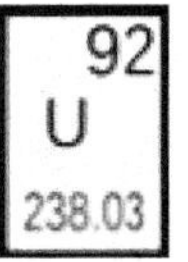

The top number is the atomic number. It tells the number of protons (and electrons) in the element. The middle letter(s) is the chemical symbol for the element. The bottom number is the atomic mass. It tells the total number of protons and neutrons in the element. In this case U238 has 146 neutrons.

Radioactivity

A common explanation for radioactivity is that radioactivity is caused when an atom, for whatever reason, seeks to lose some of its energy, in the form of atomic particles. This is called decay. (Different kinds of particles can be released.) It does this because it wants to shift from an unstable configuration to a more stable configuration. The energy that is released when the atom makes this shift is known as radioactivity. In other words, radioactivity is the act in which an atom releases radiation suddenly and spontaneously.

Some radioactivity is extremely dangerous. Some is not. The important thing to remember is that radioactivity can be measured. The unit of measure of decays per second is called the Becquerel, after Henri Becquerel, its discoverer. No one has ever been able to definitively say why radioactive elements exist.

Radioactive decay

If radioactive elements (atoms) lose particles in the decay process, what happens to the original radioactive element? Obviously, there is going to be a change that takes place. The radioactive element theoretically becomes something else. I say theoretically, because the decay process can require an immense amount of time to complete and would not be able to be observed throughout the entire process. The unstable element that is decaying is called the *parent* and the stable element into which it theoretically eventually changes is called the *daughter*.

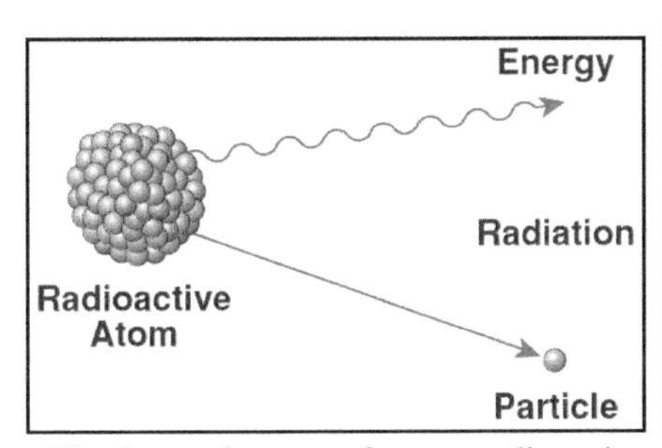

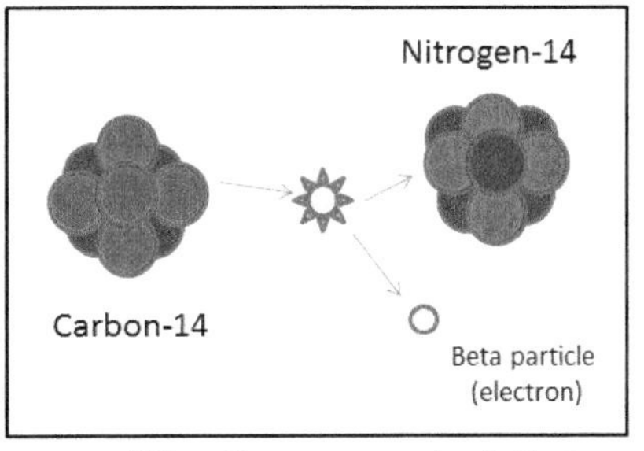

The loss of energy from a radioactive atom – The diagram on the left shows theoretically how a radioactive element loses energy. In the picture on the right, radioactive Carbon–14 is losing an atomic particle, and in so doing, theoretically becomes stable Nitrogen–14.

Many of the radioactive elements used in radiometric dating require huge amounts of time and no one has recorded or witnessed the entire decay process. For example, the time it takes for ½ of a Carbon-14 atom to decay, measured at present measurable rates is a little over 5,000 years. Then in another 5,000 years ½ of what is left decays and so on until it is all gone. Has anyone ever observed the complete process to know whether it has actually completed this process or not? How could they? Below is an illustration of the idea of half-life. Theoretically all that is left at the end of the decay process is the stable daughter element.

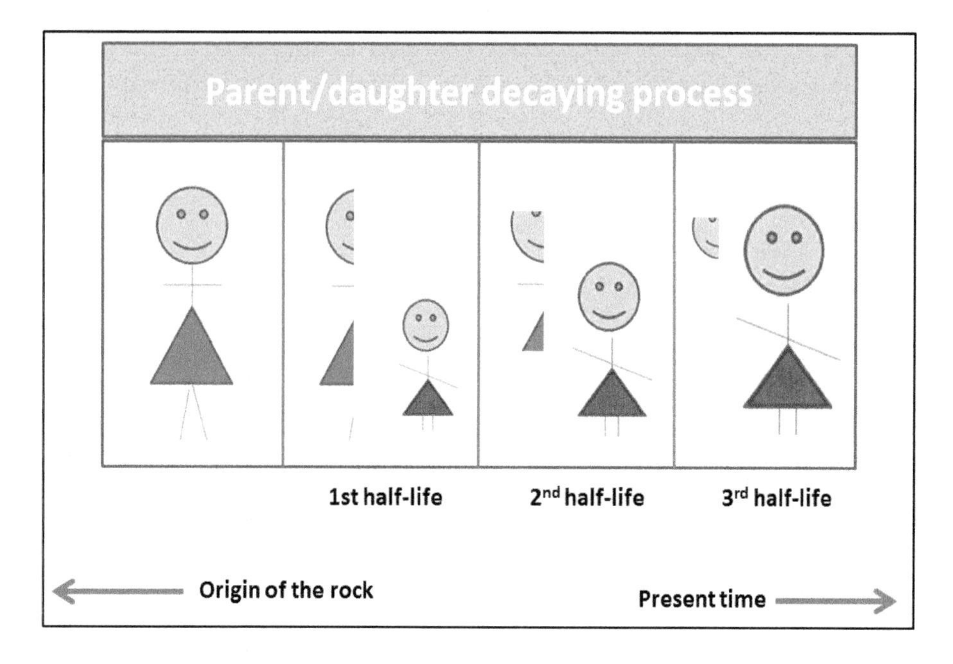

The half-life of the radioactive element is determined by the present Becquerel. The time is then projected forward or backward to arrive at a time when the process theoretically started or when it would theoretically end. In our Carbon-14 atom above, the stable element into which C-14 would decay, is Nitrogen-14. This is a chemical prediction based on what we know of present radioactive decay. Remember that the process, if totally accurate, would take thousands of years to complete. No one has been around that long.

Uranium decay

Radioactive Uranium-238 (U-238) is a very heavy element; but what I want you to notice is its long half-life from the following illustration. The question is, has anyone ever observed this? The answer is obviously, "No." Then how do we know that it operates this way? We can guess based on the chemistry of radioactive decay. It is a matter of statistics. If U-238 were to continue to lose particles like it seems to do in the present, then over a series of half-lives it will go through a chain of decays until a stable element is arrived at – in this case, Lead-206.

Uranium 238(U238)
Radioactive Decay

Type of radiation	Nuclide	Half-life
α	Uranium-238	4.47 billion years
β	Thorium-234	24.1 days
β	Protactinium-234m	1.17 minutes
α	Uranium-234	245000 years
α	Thorium-230	8000 years
α	Radium-226	1600 years
α	Radon-222	3.823 days
α	Polonium-218	3.05 minutes
β	Lead-214	26.8 minutes
β	Bismuth-214	19.7 minutes
α	Polonium-214	0.000164 seconds
β	Lead-210	22.3 years
β	Bismuth-210	5.01 days
α	Polonium-210	138.4 days
	Lead-206	stable

Uranium-238/Lead-206 is a commonly used method in calculating the radioactive decay/age of a rock. But if no one was around to actually test or record the entire process from beginning to end, then how do we know this process works? We don't! Geologists make assumptions based on the present observation of radioactivity. It is an application of the philosophy known as uniformitarianism – *the present is the key to the past.* (This philosophy was developed in the 1800's as an attempt to counteract the Bible's teaching about a global flood. We will talk more about this in Chapter 3.) So, by assuming the present has always been true in the past, geologists can use the process of radioactive decay to date rocks. How do they do this? Only rocks that are thought to have been molten at one time are considered to be valid specimens. Before we tackle this further, however, let's take a side-step and look at how fossils are dated.

Many people think that fossils are also radiometrically dated. Fossils, however, are dated by another means, also by way of uniformitarianism. Evolution says that because creatures change in the present, therefore, given enough time in the past, creatures have changed from one kind to another kind. Therefore, the rock layers with the different kinds of fossils are interpreted as evidence that evolution has taken place over long periods of time. The apparent order of the fossils in the rock layers has been assumed by geologists to be the result of an evolutionary process, having rejected the idea of a global flood.

Look at the following column. It is presented as a scientific fact. But in reality, it is a hypothetical idea.

EON	ERA	PERIOD	MILLIONS OF YEARS AGO
Phanerozoic	Cenozoic	Quaternary	1.6
		Tertiary	66
	Mesozoic	Cretaceous	138
		Jurassic	205
		Triassic	240
	Paleozoic	Permian	290
		Carboniferous	360
		Devonian	410
		Silurian	435
		Ordovician	500
		Cambrian	570
Proterozoic			2500
Archean			3800?
Pre-Archean			

The Geologic Time Table has governed all geologic dating since its development in the 1800s. If radiometric dates do not agree with this geologic bible, they are thrown out.

The column in its entirety has not been found anywhere on the Earth. Some of the rock layers do occur in an apparent order, but there is also another explanation for this apparent order that modern geologists reject. It is a general order (with many exceptions) produced by the sorting action of the Genesis Flood. If the fountains of the great deep burst open first, as Genesis states, then one would expect that marine (sea) creatures would have been the first victims of the Flood. As the Earth became inundated with water, other creatures would have followed as the Flood swallowed them up. Man, being the most mobile, would have most likely been buried last, if at all. Creatures swimming in order to survive would have ultimately drowned and their remains would have been eaten by fish or sharks.

The Geologic Column or Time Scale or Time Table

The basic Geologic Column that we have today was set by the mid to late 19th Century. The concept of time plus evolution was a philosophical shift from a Divine Creation taught in the Book of Genesis to a totally naturalistic and atheistic one taught through uniformitarianism.

As the idea of change in creatures expanded to an evolutionary view where one type of creature was thought to have changed into an entirely different creature, so the amount of time for nature to accomplish this was also necessarily expanded from several thousand years to 550 million years.

Today many people think that the Geologic Column and the millions of years it involves is a scientific fact supported by radiometric dating. Remember that radioactivity was not even discovered until after the Geologic Column had been in place for many years. Most people are unaware of this. It was not science that formulated the Geologic Column, but philosophy. It was a naturalistic attempt to explain the apparent order of fossils: that is, marine fossils on the bottom of layers, dinosaurs closer to the top. It is an idea, and an accepted idea, but not proven science.

In addition, radiometric dates are checked by the Geologic Column for accuracy and reliability. If the dates do not line up with what geologists have already determined should be the age, the radiometric dates are thrown out. Now let's go back to discuss radiometric dating of rocks.

How is a radiometric date arrived at?

Let's start with a volcanic rock. And let's say it is chemically analyzed to contain 80 parts of uranium and 20 parts of lead. Is the rock young or old? Since more of the uranium is present than lead, then the rock must be young. If the ratio was reversed, then the opposite would be true – given the following assumptions:

1. We assume that the initial state of the rock started with a certain amount of uranium and no lead. In other words, the initial state of the rock is assumed.
2. We assume that there was no lead present at the start of the process.
3. We assume that no uranium came from some other source.
4. We assume that all the lead that is present came from the decay of uranium and that it did not come from some other source.
5. We assume that the decay rate or process has not been interfered with from some other means or sources.

(In college textbooks that I have examined, these assumptions are not discussed. They are simply universally assumed to be true because of what we observe in the present.) So long as we go with these assumptions, we can statistically figure an age for the rock.

But what happens when different ages are obtained, as in the case of Mt. St. Helens lava, which was known to be ten years old at the time the dated sample was produced, but which dated in the hundreds of thousands of years? Again, geologists simply assume:

1. You or the laboratory contaminated the samples.
2. You or the laboratory made a mistake in calculations.

Mt. St. Helens recent dacite dome

But what if the sample is dated at different ages using different dating methods, as in the case of rocks from the Grand Canyon where different dates for the same rock were obtained using different methods. Then the *umpire* or *Bible* of geology is the Geological Time Scale (Geologic Time Table). Since dates back to 550 million years old were worked out in the 1800s, and have already been agreed upon, if the date for the rock appears to be too old or too young, then the dates are either thrown out or selected. The date which is the closest to what the geologist thinks it is becomes the accepted date. And that's radiometric dating.

Radiometric assumptions are not reliable

Although scientists believe that radiometric assumptions have not changed in the past, it could be shown that they have. What might have caused radioactivity to change in the past? The Genesis Flood would have added a tremendous amount of heat to rocks. It would have also added a tremendous amount of hot water to the rocks. And in fact it has been shown that radioactive decay was rapidly sped up in the past in a number of substances:

1. Carbon-14 was discovered in diamonds thought to be at least several billion years old. According to present measurements of the decay of C14, there should be absolutely no detectable C14 after around 100,000 years.
2. Carbon-14 was discovered in petrified wood thought to be a few hundred million years old.
3. Recent observed lava flows in New Zealand, Hawaii and at Mt. St. Helens dated excessively old according to modern radiometric dating methods.

Mt. Ngaruhoe in New Zealand

Recent lava flow in Hawaii

These examples show that the radioactive decay process can be affected by Earth's processes, although geologists do not know how or why yet. Flood geologists suspect that it was by way of a unique historical global geological event we call the Genesis Flood. If this is so, then thousands if not hundreds of millions of years' worth of radioactive decay occurred in a matter of a few months at some time in Earth's past. The radioactive process has been dynamically affected. This means that the radiometric clocks that geologists have developed are really not reliable. Even though radioactivity is an observable fact, the measurement and predictions of this process have been affected by something else and should not be used as a basis for age/dating. It would be like hanging on to a clock or watch that ticks, but is always gaining or losing time. Why depend on it? It's a clock, but is it reliable?

Why do geologists continue to use an unreliable method?

Geologists continue to use the radiometric dating methods in spite of the obvious flaws. Why? Although there is rarely an open public admission of these errors, the radiometric dating methods provide the only so-called *scientific* alternative to the young Earth conclusions of the Bible. At the center of the radiometric dating rational is the influence of The Enlightenment and consequently the extremely narrow view of uniformitarianism. Anything but the Bible!

These same dating methods have been used to date the various volcanic rocks in Yellowstone. So the question is, are these dates for Yellowstone reliable? Let's start looking at the geology of Yellowstone from a different framework.

Chapter Three
Interpreting the Geology of Yellowstone: The Biblical Geology of Yellowstone

In the subject of geology, it is not science vs. religion. It is religion vs. religion; Deism and/or atheism vs. Theism. The Bible makes it perfectly clear that God has been and continues to be involved in His creation. He was involved in the creating process and He has also been involved in the maintaining process. Notice this Scripture from the letter to the Hebrews 1:1-3,

> God, after He spoke long ago to the fathers in the prophets in many portions and in many ways, in these last days has spoken to us in His Son, whom He appointed heir of all things, through whom also He made the world. And He is the radiance of His glory and the exact representation of His nature, and *upholds all things* by the word of His power. When He had made purification of sins, He sat down at the right hand of the Majesty on high.... (Emphasis added.)

This is such a simple and succinct statement. The universe could not exist without His attention. This is the missing element in a uniformitarian view of the earth. Just where did the uniformitarian framework come from?

In the 1800s a particular religious belief prevailed among European intellectuals. This belief became known as Deism. Deism stated that God was involved in the creative process at one time, but does not and has not maintained or intervened in the creation since then. There are natural laws, created by the Creator, which exclusively govern the creation now and God does not have anything more to do with it. This idea effectively made God irrelevant, while still acknowledging His existence. Eventually God became just an idea, not a real personage with whom we have anything to do. This was not yet atheism, but was one step removed from Christianity and one step closer to atheism. Man still possessed an awareness of being responsible to a Creator. But if the Creator is simply an idea, then there are no real consequences for disobeying an idea. Deism is in reality a Christian heresy.

The other main attack from Deism was the disdain for the Scriptures as anything more than good moral principles. Deism ascribes things to God that are contrary to what has been clearly revealed in the Scriptures – the basis for Christian orthodoxy. The miracles in the Bible were rejected and the history contained in the Scriptures, especially the Global Flood was denied.

James Hutton, Scottish physician and *Father of Modern Geology*, and Charles Lyell, British lawyer, were two of the most influential men in the establishment of modern geology. They were also Deists. Both were highly critical of the Bible.

Deism also became the philosophical foundation of another idea, concealed in scientific terms – *uniformitarianism.* James Hutton, called the Father of Modern Geology, and a Deist, believed that the earth was much older than the Bible said. He developed the idea of uniformitarianism in his book, Theory of the Earth. Charles Lyell popularized the idea in the phrase, "The present is the key to the past." In other words, present geological processes we observe now are sufficient to explain the past history of our earth and by extension, the universe. A so-called Flood mentioned in the Scriptures was not necessary. Without stating it, the implication was clear. The invocation of a God and a universal flood to explain the past history of the earth were not required or relevant. Indeed, it was Charles Lyell's personal goal to rid geology finally and forever of any influence of the book of Genesis.

Until the 1970s, this doctrine in geology was interpreted strictly. That is, it would allow no catastrophes in the formulation of earth history. Catastrophes smacked of the Bible and that would not be permitted in science. Uniformitarianism is the foundation of modern geology. Modern geology stands, then, in the position of presenting a contrary history to the Biblical record. Both the Bible and uniformitarianism clearly tell a historical story of origins and development of man. These two stories are contradictory to one another and can never be harmonized. Modern geology will not allow a divine foot in its door. And the Scriptures clearly reject the idea of a naturalistic universe.

There is a misconception today that because creationists use the Bible (i.e. religion) that somehow that eliminates them from being scientists. What we fail to see is that uniformitarian scientists are just as religious. Uniformitarianism is a religious bias/preference. It simply chooses to eliminate God and the Bible from consideration. This is not a matter of science, but of faith.

Below is a chart that compares uniformitarian geology and Biblical geology. Both are religious. Both interpret the actual evidence within religious contexts. Both will arrive at completely different conclusions because of their preconceptions about God's involvement in His creation. This is the key. Deism was a departure from orthodox Christianity, but was not without religious influence. Both Deism and Theism are religious – one without God's involvement, the other one with it. In this context, even geology itself gets redefined. To say that geology is the study of the earth is not the whole picture. The earth is studied within a set of human biases, a framework. *It has to be this way because the remote past cannot be checked by science. We are talking about unique historical events.*

Uniformitarian Geology Deistic	**Biblical Geology Theistic**
Geology defined: the study of the earth within a uniformitarian structure; therefore everything must involve only natural processes that have been observed in the present, in order to explain the past; the scientists that subscribe to this idea are the ones considered to be authoritative on the subject.	**Geology defined**: the study of the earth within a supernatural, Biblical structure; therefore, what is observed is not necessarily what has been or what will be; human interpretation is subject to God's authority; that authority is the revelation of the Scriptures; everything must be checked against this structure.
The framework: naturalism; gradualism (also called *actualism* in some circles)	**The framework**: supernaturalism; Biblical revelation
"The present is the key to the past"	Biblical revelation is the key to the past

Man is complete and able in himself to figure out his past and future; believes in unbiased interpretation because he believes he is capable of unbiased interpretation	Man is corrupt and blind in himself to figure out his past and future; questions his interpretation because he believes he is influenced by his corruption
Present geological processes can explain the past and predict the future	Present geological processes can only explain the present
God is irrelevant in the subject of geology; the formula for research is uniformitarian framework + observation within that framework = (uniformitarian conclusions) science void of God	God is intimately involved in the geology of the earth; the formula for research is Scriptural framework + observation within that framework = (supernatural conclusions) science that involves the God who created and maintains the earth
Geology is cyclical; what will be is what is and what has been, except for a few predictable catastrophes; man's quest is to control his future	Geology consists of historical one-time events which have included God's involvement in order to fulfill His purpose; man's quest is to recognize and submit to his creator, trusting Him for the future
Fossils are evidence of past evolution	Fossils are evidence of a global flood, brought about by God and recorded in the Scriptures
The expanding universe indicates an ancient origin	The expanding universe indicates a present universe in operation as the result of a one-time creative event

The Bible presents a history that incorporates geological history

Modern geology claims to be the only scientific approach to analyzing the history of the earth and by implication, the only *true* and acceptable approach. It claims to do this through the rocks and the fossils they contain. The subject of Geology is narrowly defined as the study of the earth – *through the landforms and rock record.* In so doing it conveys a history that leaves a lot of unanswered questions. How can it be the only legitimate and true approach when it leaves out certain information that otherwise could give us a totally different historical perspective? It is my opinion that geology should not be taught as an isolated subject, but rather as an integrated whole. In other words, when examining the rocks, I need to consider *all* the possibilities and figure out which one makes the most sense. Modern geology claims to be a *history* of the earth. It therefore is comparable to the Scriptures that also present a history of the earth. But the Scriptures include many other factors that present a whole picture of history, not a disconnected history of the rocks, as modern geology does. If we only consider the disconnected history, we are going to draw a lot of wrong conclusions, which modern geology does. It leaves out much valuable information that could give us a more complete picture of history. Although the Bible is not a text specifically written to teach a *practical* geology course (oil discovery, rock and mineral identification, etc.), it nevertheless records particular historical events of our earth and the progression of life on it – from a particular perspective, a set of glasses if you will. Because it is a history record, it therefore will make statements from which geological implications can be drawn. The Scriptures are filled with such statements. Notice the following:

Genesis 1

⇨ The very first verse presupposes the preexistence of God. God created our universe and us, but He Himself existed before then. Genesis is not a history of God, but a history of Earth and man's origin. In terms of history and prehistory, there are no prehistoric times. Historic time begins when God created the heavens and the earth. If there is any prehistory, it is eternity before time and the creation we observe. There is only history – our history, and it begins with the very first verse of Genesis.

- ⇨ There are approximately 83 statements, depending on how one divides the passages, which imply significant geological events. From these we can deduce the following – 1. Physical things were created and immediately began working for the purpose for which they were created. There is no thought of gradualistic development over long periods of time. 2. The physical things were created in a very short period of time. The entire universe was created in a short period of time functioning as a working whole. This would of course include certain rocks and the biological, botanical and zoological processes of life. 3. Life was divided into kinds from the beginning. And although we don't know exactly what *kinds* means, we can see that there were boundaries among the various life entities from the beginning.

- ⇨ The creation events described in Gen. 1 actually have their completion at Genesis 2:4. Genesis 2:1 states that the heavens and the earth were *completed* and all their hosts. Everything was functioning as it was meant to from the beginning.

- ⇨ There are approximately 49 statements that record God's direct involvement in His creation. This implies that the geology of our earth was not derived naturally, but supernaturally. Such statements as, *God created, God separated, God made, God placed,* and *Let Us make* all imply that the physical things we now see were originally a product of God's work – His literal, direct work. This would include how the sun, moon and stars got there. Although the universe might show that an expansion is in progress now, since its beginning was not natural, we cannot therefore assume that the universe is like a wound-up clock. In other words, we cannot naturally extrapolate from this expanding universe that the universe is billions of years old, because it did not originate naturally. God *placed* the sun, moon and stars where He wanted them and then they immediately began functioning with the purpose for which He created them. And even though we now define laws that seem to govern our universe, the origin of the universe did not originate naturally. God is not subject to natural law.

The Geology of Genesis

⇨ Geology is the study of the earth. As we study the earth, we may be unaware that we are using a framework; a set of glasses through which we interpret the rocks and landforms around us. Nothing is interpreted in a vacuum, however. Everything we do has a context; a framework, a bias. The Scriptures present a framework in which to interpret the physical things around us. Biblical geology would follow this pattern: *knowledge of the framework (the Scriptures) + an acceptance of that framework as being legitimate + a basic knowledge and observation of the rocks and landforms + an interpretation of those things based on the Scriptural framework = balanced and correct geology*. It is interesting to note here that in Genesis 2:15-25 the same type of framework is implied: God originally created. This is the framework.

⇨ Man was given certain things to do under God's authority. This is the acceptance of that framework. Man is under God's authority.

⇨ The animals and birds were brought by God to the man to see what he would call them. This is the interpretation of the created things in light of that framework. In short it is interaction with the creation *within a God-centered framework*. That is essentially what the modern creationist movement is all about.

Secular geologists and flood geologists see the same evidence! Except for the time frame involved, the evidence is the same! Here is a list of the geologic phenomena that both sets of scientists agree on:

⇨ Mountain uplift; mountain building
⇨ Rifts, faults and gashes
⇨ Sudden appearance of life in the Cambrian layers
⇨ Erosion
⇨ Sea coverage; marine deposition; abundance of marine fossils
⇨ Sediments – some of them extremely thick
⇨ Hydrothermal alteration
⇨ Radioactivity
⇨ Tectonic activity; earthquakes
⇨ Sea floor eruption of basalt

⇨ Extinction of many groups of plants and animals
⇨ Variation and Mutation
⇨ Volcanism
⇨ Past glacial activity
⇨ Rock layers; sedimentary rock layers containing an abundance of fossils
⇨ Disarticulated fossil bones; dinosaur fossils

How can geologists see the same geologic evidence and yet be so far apart on its interpretation? The issue has been and will always be worldview.

Genesis 2

⇨ There are at least 18 passages that reveal God's direct involvement in creating man and woman. These are indicated by phrases such as, *God had not sent rain, God formed man, God breathed, God planted, God placed the man, God caused to grow, God put the man into the garden, I will make him a helper, God formed, God brought, God caused, God took, God closed up, God fashioned, He had taken, God brought the woman to the man.* And although this degree of involvement is sometimes not apparent throughout Biblical history, it nevertheless does take place and is recorded in the pages of Scripture. These totally contradict the religion of Deism which would state that God is no longer involved in His creation. With this in mind, the assumptions formed within the framework of Deism, i.e., Hutton's Theory of the Earth are therefore non-Christian. And anything that flows from this line of reasoning, i.e., uniformitarianism, is also not Orthodox Christian belief.

⇨ There are at least 15 passages that indicate significant geological and biological one-time events:

1. No plant of the field had yet sprouted.
2. God had not sent rain on the earth.
3. God formed man of dust from the ground.
4. Out of the ground the Lord God formed every beast of the field and every bird of the sky.
5. Man became a living being.

6. The man gave names to all the cattle, and to the birds…and to every beast of the field. (This implies cognizance right from the beginning. This is not gradualism, but well-formed intelligence exercised from the beginning.)
7. Morality and a consequence of disobedience – introduction of death.
8. Woman formed from a rib taken from Adam.
9. The reason for man leaving his family unit – to unite with his wife; marriage.

Uniformitarian evolution states that man originally came from an apelike creature, which in turn, came from some other life form, etc. But from the above passages of Scripture, it is clear that man's history did not involve this. He originally came from the dust. God then formed him, breathed life into him and he became a living being. All this is summed up in Gen. 1:26 – "Let Us make man in our image…male and female…."

Some evolutionists think that woman evolved differently than man. The uniformitarian viewpoint will only allow a naturalistic explanation for man's rise and therefore will naturally take an opposite viewpoint to the Scripture. At the heart of uniformitarianism is a man-created bias against God. The Scriptural viewpoint teaches that man and woman are special in the panoply of creation. They were uniquely created in the image of God. This is the reason why we have laws protecting human beings. The only reason for doing so is that man was created in the image of God. There is something special about man. If he had been derived naturally just like all the other animals, as evolution teaches, there would be absolutely no reason for treating him differently than any other creature or creation.

As geology is the study of the earth, that would involve the study of the past history of the earth, the present processes of the earth and the future of the earth. But, in what framework will one study the earth?

In establishing a *whole* Biblical framework, it is extremely important that we integrate these various dimensions. Modern geology attempts to limit its study and scope to physical properties and processes of the earth. This is because it is only interested in the study of the natural world,
claiming that this is the only scientific view and therefore the only right view, or at least the only permissible view. But in so doing, it skews the development of a complete worldview. By omitting many things that otherwise would be included in the study of the earth, it *a priori* establishes

a naturalistic and therefore an atheistic worldview – by default.

There are many references in the New Testament that speak to the study of the earth. Integrating these references will help develop a whole or complete view of Genesis.

In 1 Peter 1 of the New Testament there are at least 18 references to earth history study. Here are just a few:

- ⇨ 1 Peter 1: 3, 21 reference the "raising of Christ from the dead" by God. This is a long way from a Deistic view, which would deny the resurrection. God was clearly involved when Christ rose from the dead, a historical fact that many witnessed. If God intervened in the history of the earth to raise Christ from the dead, what might this say to the history and future of our globe?
- ⇨ In the study of the earth, one will readily notice that things wear out, grow old and perish. In contrast to this observation, verse 4 states that the inheritance that Christians have received is imperishable…will not fade away. If this is so, what might this say about the history and future of our globe?
- ⇨ In the study of the earth, time is an unmistakable part of the processes of our globe. It is interesting that James Hutton, framer of the uniformitarian view of earth history, saw earth history as cyclical – without beginning or ending. It is difficult to imagine time as being cyclical. Time by its very definition implies a beginning. In sharp contrast to this, verse 5 teaches that earth history will eventually come to an end, "…in the last time." If this is so, what might this say about the history and future of our globe?
- ⇨ In a uniformitarian/Deistic framework of earth study, the physical/material is all there is and is all that really matters. In this kind of thinking at the very least, God does not matter. But in verse 12 notice that the physical/material world is not all there is – "…these things which now have been announced to you through those who preached the gospel to you by the Holy Spirit sent from heaven – things into which angels long to look." If this is so, what might this say about the history and future of our globe?
- ⇨ In the uniformitarian framework of earth history, morality is irrelevant. Morality is a man-made thing and therefore is only expedient or non-expedient as determined by the individual. But notice 1 Peter 1: 15-16 says, "…but like the Holy One who called you, be holy yourselves also in all your behavior; because it is written, you shall be holy, for I am holy." Morality exists because God exists

and is holy. The reason we should act in such a way is because God is that way. If this is so, what might this say about the history and future of our globe?

⇨ In a uniformitarian framework of earth history, the subject of eternity (forever) is a foreign subject. Modern geology is only concerned with what we can observe now as it relates to the past history of the earth.

Interestingly modern geology violates this premise by making bold and sweeping statements about the unobservable past, as if scientists had actually been present to observe them. As an example, take this statement from Extreme Science.

> In the very beginning of earth's history, this planet was a giant, red hot, roiling, boiling sea of molten rock - a magma ocean. The heat had been generated by the repeated high-speed collisions of much smaller bodies of space rocks that continually clumped together as they collided to form this planet. As the collisions tapered off the earth began to cool, forming a thin crust on its surface. As the cooling continued, water vapor began to escape and condense in the earth's early atmosphere. Clouds formed and storms raged, raining more and more water down on the primitive earth, cooling the surface further until it was flooded with water, forming the seas.[1] (Emphasis added)

In 1 Peter 1:24-25, in contrast to the words or pronouncements of science, which by definition is not absolute because it only deals with things that can be tested or observed, the words that God has revealed are eternal and therefore absolute.

Genesis 3

In Genesis 3, there are **18** specific references to God's direct involvement in man's affairs. This has been consistent from the beginning. The Bible teaches the opposite of a Deistic concept of God. Missing this has

[1] http://www.extremescience.com/earth.htm

promoted an atheistic framework in which modern geology has developed. And because of this framework, the reasons for morality, consequences of immorality, the origin of evil and decay are totally missed.
One of the liabilities present in rejecting a literal Genesis is the difficulty in explaining man's personality and moral choices. His constant wrestling with right and wrong cannot be explained within the framework of uniformitarianism. But a straightforward reading of Genesis chapter three gives a satisfying and thorough answer. Uniformitarian geology might be able to explain earth's rocks and landforms, albeit with great difficulty, but it cannot help us humans who endeavor to answer the questions of whom we are and from where we came.

Within the Genesis framework, there are at least **30** references to a *moral/immoral history* that took place early in man's creation. Notice just a few of the following examples:

- ⇨ The introduction of a personage who undermines God's authority in His creation.
- ⇨ The introduction of certain aspects of created man such as communication, reasoning ability, the ability to make choices, consequences for disobedience and loss of happiness and freedom.
- ⇨ Geological, environmental, biological and emotional consequences for disobeying the Creator. Notice that Romans 8:20-22 refers to these same consequences. Man's disobedience brought on the decay in our present world. Man's disobedience brought about the Flood which destroyed the original created landforms. Man's disobedience brought about the enmity between man and man, between man and Satan and between man and God. When we either diminish or disregard Genesis, we cannot find a satisfying and reasonable explanation for the decay and disharmony we observe in the world today.
- ⇨ Even after man and woman sinned, God is still involved in their care for at least the things that they do not anticipate (Gen. 3:23-24).

Other things to notice from Genesis 3 are the significant changes that came about in man and in his environment as a result of his disobedience:

- ⇨ Gen. 3: 1, skepticism, doubt and questioning of God's authority and an undermining of His authority
- ⇨ Gen. 3:1, for some reason Eve was completely unaware of what was happening to her. Here is one of the creatures that the very God she trusted had created. She evidently did not suspect the serpent's motives. One of the most destructive things brought into God's

beautiful creation, then, was the introduction of evil – not by God or man, but from someone who had turned evil. This is one of the greatest mysteries in all of Biblical history. One could ask, "Why?" all day long and never really understand God's plans. Charles Darwin's own daughter was physically weak. She died while yet young. Darwin blamed God for her death. Ironically, Darwin could not explain her death and therefore concluded there was no God. His god effectively became the very thing that he hated – the random, senseless, process of evolution with its survival of the fittest and destruction of the weak. Turning away from God will not provide a satisfactory answer. It will only make the picture worse.

- ⇨ Gen. 3:2, unhealthy reasoning about God's word and His authority
- ⇨ Gen. 3:6, a moving away from trust in what God says to what man says or reasons – "When the woman saw…"
- ⇨ Gen. 3:6, disobedience to what God has clearly spoken and tempting others to do the same
- ⇨ Gen. 3:7, an unhealthy opening of the eyes; enlightenment as the result of disobedience. This incident in Genesis 3 is the exact pattern followed during another historical enlightenment, The Enlightenment or The Age of Reason in which eyes were opened to a new geology. This new geology has been responsible for creating the environment of unbelief and skepticism that has dominated science for the past 200 years. Much damage has resulted including the various deistic religions, the reinterpretation of Scripture, the schisms that exist in the modern church, eugenics, racism, Nazism, and communism.
- ⇨ Gen. 3:8, an unwholesome fear of God: being afraid to be exposed to Him, hiding from Him
- ⇨ Gen. 3:13, blame-shifting for one's disobedience
- ⇨ Gen. 3:13, deception in a once peaceful and orderly creation
- ⇨ Gen. 3:15, spiritual warfare between the serpent and man – as long as the serpent exists. (In Revelation 20 we see what life will be like when the serpent is bound in the abyss for 1000 years.) Because of this enmity mentioned in Gen. 3, there will be problems in God's once perfect creation between man and man, and man and woman, all brought about through the serpent's deceptive lies about God and His word, and man's disobedience.
- ⇨ Gen. 3:16-19, whatever else these verses may mean, they imply great difficulty in life and then comes death
- ⇨ Gen. 3:17 tells us of a connection of man's rebellion and its effect on the earth itself. The ground was cursed. I wonder if there might be a correlation between this and the Law of Entropy, which says

that the earth is wearing down and growing old. This Law was discovered around 100 years after Hutton's idea of a cyclical history for the earth. He could see no apparent neither beginning nor ending for earth history. This followed on the heels of a rejection of the Book of Genesis. The Law of Entropy rather validates what Genesis records – the ground was cursed.

- ⇨ Gen. 3:1, 20 imply that Adam and Eve were the first humans; she was the first woman with whom the serpent dialogued after creation and Adam calls her the mother of all the living – humans, that is. Also implied here is the exalted view that Adam had of man – God's creation. If he was just like all the animals as a uniformitarian framework teaches, Adam would have been arrogant for calling his wife Eve. But he knew he and Eve were specially created by God different from all the other animals – "She is the mother of all the living."
- ⇨ Gen. 3:22, one of the biggest changes that came about because of man's disobedience was the establishment of a bent to follow his own will. By an act of will he disobeyed God. And now because of his *enlightenment*, he would continue to exercise his own will – but with a knowledge of good and evil: potentially disastrous consequences for man's future. What would have happened if in that condition, he had eaten from the tree of life? I think the implication is that man would have produced rebellious men and women FOREVER. There would have been no conclusion or ending to the mess.

The rest of Scripture shifts focus from the garden and man as the special creation to the establishment and fulfillment of a promised savior who would save man from his dilemma and bring in a new heaven and earth. THAT is the conclusion to the mess brought about in Genesis chapter 3. And it ends with Revelation in which evil and the serpent are finally judged, man is eternally saved, and God is once again glorified and the head of His creation. This is one of the main reasons why it is important to teach a Scriptural view of geology. The creation, fall and flood supply the whole foundation for a bright future for man. A uniformitarian view only supplies a foundation of nothingness and meaninglessness.

Genesis 4

In Genesis 4 some obvious changes seem to have taken place since the creation of the heavens and earth and the fall of man. **1.** The amount of interaction between God and man has apparently diminished – only seven instances of God's direct involvement. **2.** Man increasingly becomes more self-centered. This is seen in Cain's concern for his own safety rather than a deep sorrow for the murder of his brother. He names a city after one of his sons. Why not after the merciful God who has protected his life? Within the fifth generation after Cain, Lamech has taken two wives. This was deviation from God's plan in chapter two. He displays arrogance about his sin of murder. "If Cain is avenged sevenfold, then Lamech seventy-sevenfold." **3.** There seems to have been a drifting from the Lord for a while until Seth comes along. Verse 26 states that with Enosh, Seth's son, men began to call upon the name of the Lord. This sets up two lines of response toward God. Man either calls upon the true God or he takes other paths that lead away from the true God.

Genesis 4 is a fulfillment of what God had told the serpent, "I will put enmity between you…and her seed…" in chapter three. The enmity is seen in the struggle that Cain has with Abel. It is seen in the struggle to either live for one's self, as with Cain and Lamech, or to call upon the name of the Lord, as with Enosh.

Genesis 4 also introduces the concept of rightful worship. The word **offering** and the phrase, **call upon the name of the Lord**, show up for the first time. Uniformitarian anthropology teaches that man evolved in his act of worship out of a need to explain the world around him. Genesis teaches that man brought offerings to and called upon the name of the Lord out of his responsibility and duty to his creator. Corruptions of this responsibility followed with the various mythologies and paganism born out of man's rebellion and willful ignorance of his creator. Chapter four begins to display the consequences for Adam and Eve's original disobedience to God through the acceptance of false religion.

Another first appears in Genesis 4: the correlation between man's emotional well-being and his struggle with sin. God told Cain, "If you do well, will not your countenance be lifted up? And if you do not do well, sin is crouching at the door; and its desire is for you, but you must master it." The Scripture teaches that personal responsibility for my emotional well-being lies with me. Our world today often teaches that my emotional well-being is first dependent on my environment, not on my relationship to God and my ultimate responsibility to Him.

From the Genesis 1-4 we see the context for a Biblical history and why certain things are recorded.

1. God created man.
2. Man rebelled against his creator.
3. The struggle between man and the serpent is set.
4. God promises a savior and the death of the serpent.

The rest of the Scriptures will record how God fulfills this promise and how man continues to respond to his maker. This is the context for a Biblical framework.

Genesis 1-4 certainly provide logical answers to many questions. A Biblical view of geology can answer the following conundrums:

- ⇨ Why there is apparent order to the universe
- ⇨ Why our planet seems to be the only planet suitable for life
- ⇨ Why there are biological boundaries among the plants and animals
- ⇨ Why our universe holds together rather than fly apart
- ⇨ Why man mistreats his fellow man and his creation around him
- ⇨ Why man lives selfishly
- ⇨ Why our planet seems to be running down
- ⇨ Why man has emotional problems
- ⇨ Why there is a need for man to worship his maker
- ⇨ Why there is marriage
- ⇨ Why man struggles with morality
- ⇨ Why there is a contradiction between the idea of a loving God and the existence of evil
- ⇨ Why lies and deceit exist
- ⇨ Why there exists hope in spite of the decay all around us
- ⇨ Why man knows that things in this world are not quite right

At this point, someone may ask, "Can't Genesis be harmonized with what scientists say is a product of millions of years of evolution?" Below is a chart of comparison between a uniformitarian view of earth history and the one taught in Genesis:

Uniform geological processes; no apparent beginning and no apparent ending to naturalistic processes	The very words, "In the beginning…" state there was a beginning.
The universe was the result of a *Big Bang* which then began to spread out over at least 15 billion years. In this scenario, the earth would have come much later at around 4.6 billion years ago.	Genesis states that first came the space into which the planets, sun and stars would be placed, after the earth had been created on Day One of creation. The sun, moon and stars were created after the earth.
There was a gradual spreading out of the universe. This was a random and purely naturalistic process.	God places the sun, moon and stars where He wanted them, according to His plan. There was nothing naturalistic about it.

Genesis 5

Genesis 5 observations:

⇨ It is clear that this chapter is not a collection of random recordings. It is meant to record and make known a specific history associated with Adam. During the 130 years after Adam's creation, we know he had other sons and daughters. Some of those mentioned are Cain, Abel, and the wife of Cain. Some of the people who built the city along with Cain were probably also Adam's direct descendants. But Seth came after all this. Why does the Scripture pick up with Seth and state, "This is the book of the generations of Adam," when in fact there were others before Seth? I believe the answer lies in Gen. 5:21-24, 29. "Enoch walked with God." The connection to man's God is clear with Enoch, Lamech, and Noah. The interest is in recording the specific genealogy of Adam through the children who worshiped the true God. This Noah, who, according to Lamech, "…would give us rest from the toil of our hands arising from the ground which the Lord has cursed," (Gen. 5:29) would continue to establish this genealogy down to Abraham, through whom the Messiah would come.

- ⇨ A word about the supposed gaps in the genealogy in chapter five – some will say that this genealogy was not meant to be a chronology, but simply a genealogy. The part of this statement that is true is that chapter five records a genealogy. But what is the point of a genealogy if there are gaps in the genealogical *time* sequence? This argument was made some time ago to accommodate the long ages postulated by uniformitarianism. There are three points to consider when examining the genealogies in Genesis chapter 5.

  1. The age of the father is given when he had the son that was recorded.
  2. The number of years the father lived after he became the father of the son that is recorded. And
  3. The total number of years the father lived. So, not only is there genealogy in this passage, but also *chronology*.

Let's think about this for a minute. If there were significant time gaps between descendants, say of several thousand years, how could one be sure that they had a reliable genealogy? When establishing a genealogy, it is important to establish step-by-step descendants. What is the purpose in telling you that I am a descendant of William the Conqueror when I cannot show how I am his descendant? The evident purpose in chapter five is:

- ⇨ To show that Noah was a descendent of Adam – from one descendent to another all the way back to Adam.
- ⇨ To show the amount of time that had elapsed between Adam's creation and the man called Noah. Otherwise we have no genealogy. This same pattern is followed all the way through the Scriptures. Take the genealogy of Jesus in the New Testament for example. It was important for Jesus to prove His pedigree from himself to Abraham, as the promised seed of Abraham, and then to Adam to establish himself as the promised seed of the woman, Eve, who would defeat the serpent. Otherwise anyone could claim to be the promised savior.

It would seem then, that unless otherwise stated, genealogy and chronology go together. The only gap in chapter five is the 130 years between Adam's creation and the birth of Seth. Adam did have other children, but God chose to record the genealogy of Adam through Seth. And it is his set of relatives that the Scripture follows. There are no gaps in Adam's genealogy *through* Seth. The only gaps are between Adam and his other descendants. But those aren't the concern of Genesis.

Earlier I had stated that modern geology is a history just as the Bible is a history. The philosophy behind modern geology is uniformitarianism. It teaches that man's genealogy lies somewhere in the remote past through millions of descendants. This is accepted and taught as fact, even though that genealogy has never been proven through the various fossils of descent that would be needed to establish such a claim. The Bible also presents a genealogy of man. And this genealogy is recorded, son-by-son, and father by father, all the way back to the beginning. But its history is quite a bit different than uniformitarianism. Uniformitarianism is therefore in reality a diversion from orthodox Christianity. We must not let the scientific language fool us. Uniformitarianism presents a totally different and contradictory history of the origin of the universe and man. It claims to reveal the true history, or at least the only acceptable one, scientifically. But underneath all the scientific verbiage is a competing philosophy that leads to disaster for mankind: an alternate path of truth as the serpent promised Eve.

Genesis 6

Genesis 6 gives us the reason for the global flood in Genesis 7-9. When we interpret Genesis within a naturalistic framework, we take away

- ⇨ God's direct involvement in His creation
- ⇨ Cause and effect for sin
- ⇨ Man's responsibility to his creator
- ⇨ Moral consequences for disobedience and rebellion

The Scriptures will not make sense in a naturalistic/uniformitarian framework. Adam and Eve become a *symbol* for the universality of man, the serpent becomes a *symbol* for evil, and sin becomes an evolutionary quirk, an imperfection, not a moral breach. The universal flood becomes a local flood and the ark becomes a *symbol* for Noah's attempt to explain his religious feelings.

Within a supernatural framework, however, chapter 6 becomes an extremely important chapter as it highlights man's **responsibility** to His creator and **consequences** for his rebellion. The Flood was necessary because man had become so corrupt that it was ruining the earth. Jesus later uses this historical event to warn people of another judgment coming for the same reason. The history of the Flood becomes a critical piece of the truth that God holds man responsible for what he does and says. If you take away the *reason* for the universal Flood, you take away the universal Flood.

Genesis 7

This is the Flood Chapter.

⇨ Noah built, gathered animals into and entered the ark at the command of God

⇨ Noah and his family were the only humans on earth who had found favor with God

⇨ The words ***all*** and ***every*** (16 times in this chapter alone) indicated that the flood was global in its extent

⇨ Noah was 600 years old when the flood came on the earth – indicates a unique and specific flood; as indicated earlier, this was not just a natural event. The flood was brought upon the earth by God as judgment. He was directly involved in it, not a passive naturalist.

⇨ Living things in the Bible apparently were divided according to what lived on the face of the earth: **(1)** mankind, **(2)** animals, **(3)** birds, **(4)** creeping and swarming things, **(5)** beasts, and **(6)** cattle. Marine life would have been excluded.

⇨ The geology of the flood: **(1)** the fountains of the great deep would include reserves of water, steam, and reserves of minerals, **(2)** the phrase, "burst open" would include earthquakes, cracks, making of metamorphic rocks, volcanic release **(3)** tidal waves, **(4)** churning and releasing great amounts of sediments, **(5)** all the high mountains were covered by 22.5 feet, **(6)** all land-dwelling animals would have been destroyed; some would have been instantly buried, **(7)** The earth stayed covered for almost a half year. Lots of settling of layers of sediment would have been created.

⇨ The floodwaters receded off the earth: There would have been a tremendous amount of sheet and channelized erosion with huge amounts of water rushing off the earth. A further note on this is given in Psalm 104:6-9 where we read about rapid mountain building. The uplift of mountains would have produced a tremendous amount of heat and pressure and therefore metamorphic rocks.

⇨ The post-flood world would have been drastically altered from the pre-flood world. Nothing would have been recognizable to Noah. No landmarks would have been preserved. The only thing that would not have changed would have been East and West, as marked by the sun. Noah would not have been able to recognize where home used to be. Everything would have changed.

It is within the Biblical framework of a young earth and global flood that Genesis presents its geology. If one starts with the premise that Adam was the first human, as Genesis states, then the time frame from Adam to Noah is roughly 1,600 years. This makes the ancient ages given by geologists for Yellowstone untenable. Therefore, there has to be something wrong with the radiometric dating system as it is presently used.

It is also within the Biblical framework of a young earth and global flood that we will examine the geology of Yellowstone National Park. Charles Lyell proposed the notion that earth history was molded through gradual, lengthy geologic processes. And he formulated this idea within the context of bias against anything to do with the Book of Genesis and especially a global Flood. His personal goal was to eliminate any influence of Moses from geology. Every secularly trained geologist since has followed in his footsteps. And since that time, Yellowstone geology has been formulated within that framework. Those of us that believe the Scriptures need to re-evaluate this position.

Yellowstone and the Genesis Flood

Is there any geologic evidence in the Yellowstone region to support the Bible's claim of a global flood about 4,500 years ago? There is quite a bit of evidence!

⇨ First and foremost, there is the Yellowstone super volcano itself. Not since this eruption has there been anything this large. As a matter of fact, the geological record clearly indicates that the frequency and size of volcanic eruptions have been steadily growing fewer over time. If uniformitarianism is a true historical account of earth's history, then the volcanic eruptions should be about the same as those of today. After all that is what secular geology teaches – *"The present is the key to the past."* But the geological evidence gives a picture of greater volcanic episodes in the past. The idea of a global flood would certainly fit this evidence.

⇨ The unbelievable size of the Yellowstone caldera is jaw-dropping. No one has witnessed such a huge eruption in known history. The crater is 45 miles long and 35 miles wide. And this was not the largest of the Yellowstone eruptions either. The largest crater is 75 miles long!

A portion of the Yellowstone Caldera Rim can clearly be seen in the background of this picture.

⇨ The amount of tectonic uplift and planed surfaces speak of the recession period of the Flood after the floodwaters had reached their maximum at day 150 of the Flood. At that point, according to Psalm 104:5-9, the mountains rose and the waters began to rush off in sheets and then in channels, leaving huge erosion features in the aftermath.

It is very interesting that there are planed surfaces at 11,000 feet in the Beartooth Mountains, just northeast of Yellowstone. What kind of geological forces would plane these mountains? The recessive stage of the Flood provides a good explanation.

The planed surfaces of the Beartooth Mountains.

Lone Peak (also called Lone Mountain) at Big Sky, Montana west of Yellowstone National Park: an ugly 11,166 feet high volcanic remnant. In the winter, it is a popular ski resort. Why is it so ugly? Geologically it is called an *intrusion*. It is mostly built of andesitic magma that pushed itself upwards during the Flood through freshly laid sediments. It had been covered by sedimentary layers and so never erupted as a full-fledged volcano. As the Floodwater washed off the overlying layers, during the recessive stage of the Flood, the massive mountain was exposed.

Up toward the summit of Lone Peak there are very interesting sedimentary layers that have undergone extensive folding caused by uplift; again, evidence of the power and extent of the Flood.

⇨ Massive limestone formations and folded sedimentary rock – what geologic force would bend rocks? Rocks don't bend, they break. The only explanation of folded sedimentary rocks is that they were quickly folded shortly after they had been deposited, as in the case of the rocks around the Lewis and Clark limestone region, northwest of Yellowstone.

The massive amount of fossil limestone in the Lewis and Clark region shout of a cataclysmic water event, not slow and gradual deposition of limy, shallow sea deposits over millions of years.

Folded sedimentary rocks near the Lewis and Clark Caverns State Park, Montana, northwest of Yellowstone.

Beartooth Butte near Top-of the-World store in the Beartooth Mountains is a real paradox. Resting directly on top of basement rocks is a butte of sedimentary rock containing all kinds of marine fossils. In the secular geological scheme of things an entire geologic period is missing – the Silurian! 20-25 million years is missing and there is no evidence that it went missing. Also, the contact between the Butte and the basement rock is flat and it represents close to 2 billion years of missing geologic history.

And in fact the entire Silurian Period is missing in all of Yellowstone! And where is the rest of the sedimentary rock that obviously extended for a much greater distance than what has survived in the Butte? Certainly there is a better story for how this lone butte at around 10,000 feet above sea level, high in the Beartooth Mountains, came about. There is! Very simply the chronology might look like this:

1. The Genesis Flood laid down the sedimentary deposits directly over the basement rock at a much lower altitude.
2. Then the mountains rose, according to Psalm 104:5-9, and the waters started to rush off. The Beartooth Mountains were raised and the Flood waters took most of the sedimentary layers that had been freshly laid down in the Flood and carried them away.
3. Later the Ice Age continued the erosion and carried more away with glacial movement and flooding. Today, the remnant known as Beartooth Butte survives. It is a very simple and yet profound explanation.

Beartooth Butte

Much of the limestone in and around Yellowstone is called Madison Limestone. It is a huge formation and covers a lot of the Park surrounding region. The picture above and those on the next page show the extensive limestone deposits around Big Sky and the same formation near Red Lodge, Montana. Such huge amounts of limestone cannot be the work of slow and gradual deposition over millions of years, but the work of some larger catastrophe – the Genesis Flood.

Extensive limestone deposits near Big Sky, Montana

The Madison Limestone near Red Lodge, Montana

A massive limestone mountain in the northeast part of Yellowstone Park

An amazing formation occurs along US Highway 191 of the Gallatin Canyon. A huge wall exposed along the highway between Big Sky and West Yellowstone is made of limestone and is over 300 feet long and at least 40-50 feet high. It is filled with calcite vugs of nicely formed dogtooth calcite. A vug is a cavity generally lined with crystal, typically quartz or calcite. Check out the following pictures.

Dogtooth calcite crystals inside a vug

Another great piece of evidence for the Genesis Flood in the Yellowstone area is the presence of fossil ripple marks in the Gallatin Mountains above Big Sky, Montana - 8,500 feet above sea level! These are scattered all over the mountains in the area. They are the rule, not the exception.

Fossil ripple marks at 8,500 feet above sea level

Mt. Everts is a strange feature in Yellowstone Park. It sits opposite of the active volcanic area of Mammoth Hot Springs. It is a sedimentary mountain containing a huge amount of land and marine fossils. This was one of those leftovers as the Flood receded. It is not surprising that it is sedimentary, because Mammoth Hot Springs is travertine terraces, formed from hot spring water that has moved through limestone underneath. What is surprising is the unconformity that exists at the top of Mt. Everts. It is missing lots of geologic time.

Mt. Everts near Mammoth Hot Springs

The limestone is supposed to be Cretaceous sedimentary sediments approaching 100 million years old. But in the next picture, notice the small chunk of rock at the top on the right-hand side. It is volcanic and from the pyroclastic tuff flows of the Yellowstone Caldera supposed to be 600,000 years old. They rest ***directly*** on top of the Cretaceous sediments. This is called an angular unconformity in secular geology and shows that there is missing time of almost 100 million years! Furthermore, secular geologists point out the discoloration where the tuff rests directly on top of the sedimentary sediments. This is where the hot pyroclastic tuff literally burned the sedimentary sediments! In other words, there is no missing time. The Genesis Flood would explain that the two events happened sequentially, rapidly and in quick succession of each other.

The top tip of Mt. Everts is the Huckleberry Ridge Tuff from the volcanic eruption of Yellowstone supposedly that occurred 600,000 years ago. It rests directly on top of Cretaceous sedimentary deposits, supposed to be anywhere from 146-65 million years old.

Another phenomenon of Yellowstone is the great number of petrified trees in the Gallatin and Absaroka Mountains – hundreds of square miles of petrified trees. These had historically been interpreted as 'petrified forests' meaning that successive forests had come and gone over thousands of years. But then in 1980 that was all changed by the eruption of Mt. St. Helens. When it was observed that the mud flows had ripped up and then deposited trees in all kinds of positions, the so called *petrified forests* were reinterpreted to be more in line with the catastrophic burial of thousands of trees that occurred at Mt. St. Helens.

This is one of the few easily accessible petrified trees in Yellowstone Park. Tourists have ravaged the others over time. The tree appears to be in growth

location. But it isn't. It is in growth *position*, having been moved there by water and volcanic mudflows. Later as the sediments were removed by erosion, the tree became exposed; giving the appearance of a forest that once grew there. The fact that it does not have roots shows that it did not grow here.

The geology of Yellowstone is indeed complex featuring tectonic mountain building, earthquakes, several volcanic eruptive events, pyroclastic flows, Ice Age sculpting and catastrophic Ice Age flooding. But it can all be explained by interpreting these geologic wonders using the framework of the historical Scriptures. The forces of modern secular geology are strong and intimidating, intellectually threatening anyone who embraces the Bible in interpreting earth history. But time and again the Bible has demonstrated its veracity by better explaining the geological phenomena of our natural world.

It was this book, **Principles of Geology, Volume 1,** that Darwin took on board the H.M.S. Beagle. It was to prove to be monumental in the formation of his worldview. Darwin later claimed that it was Lyell's book and his discussions of uniformitarian geology that give him the time needed for his biological evolution.

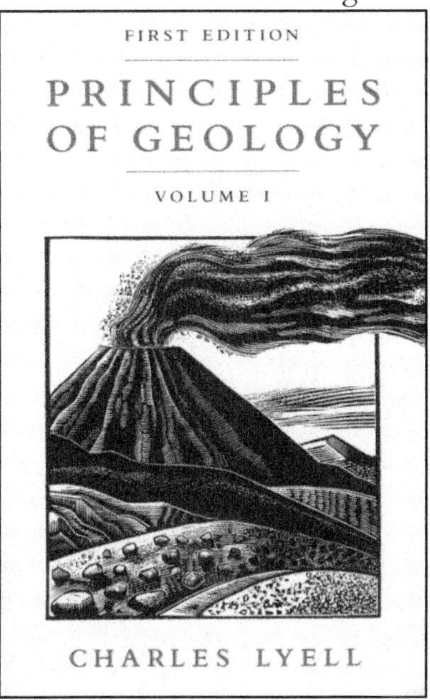

Chapter Four
The Yellowstone Super Volcano

A super volcano is any volcano capable of producing a volcanic eruption of ejected volcanic material with a volume greater than 240 cubic miles. This is thousands of times larger than normal volcanic eruptions. Yellowstone's eruptions have certainly exceeded that! By comparison Mt. St. Helens ejected about .25 cubic miles of volcanic material.

Geologists believe that Yellowstone is actually the result of several related volcanic eruptions throughout history, originating 16.1 million years ago along the Oregon/Nevada border, and culminating in the most recent (and best known) eruption 600,000 years ago. Of course, I do not endorse the ages of these eruptions for two reasons. **(1)** They conflict with the Biblical worldview, and **(2)** They are based on the many improvable assumptions involved in radiometric dating. What I do find interesting is that these eruptions all took place, in modern geologic terms, very recently in earth's geologic history. I believe that this actually places them within the Genesis catastrophic global flood event.

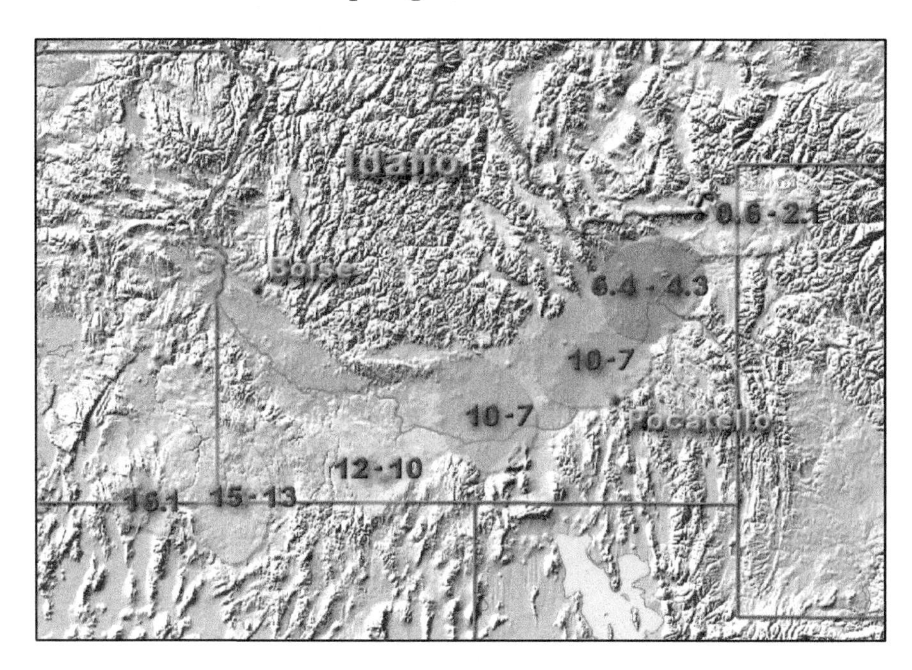

The volcanic evidence for these super-sized eruptions is real. The dates are derived from radiometric dating discussed in chapter 2. If we take away the questionable radiometric dates, the picture that is left is one of catastrophism caused by the great Genesis Flood.

Together the last three eruptions of Yellowstone, ejected more than 875 cubic miles of volcanic material, 3,500 times the amount that Mt. St. Helens ejected in 1980! There seems to have been another caldera eruption, minor in comparison, that geologists date around 174,000 years ago, which formed the West Thumb of what is now Yellowstone Lake. (We will look at that later.) If we ignore the unreliable radiometric dates for a moment and simply consider the amount of volcanic material ejected from these eruptions, we cannot help but consider them catastrophic events unlike anything that has happened in known recorded history.

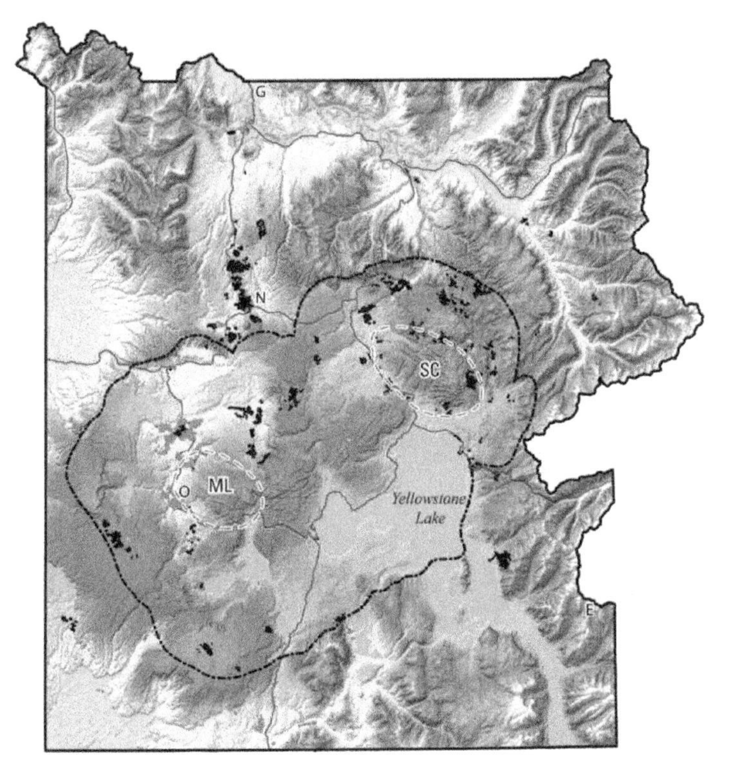

Geologists have traced what they think is the boundary of the Yellowstone Caldera. It stretches approximately 45 miles long and 35 miles wide – an incredibly large volcano!

What force would have caused such catastrophic volcanic eruptions? There is nothing going on today that would even remotely provide us with a framework in which to interpret these huge eruptions. The present modern geologic framework of uniformitarianism (*the present is the key to the past*), simply will not work to explain these massive volcanic events. We must look for some other interpretive device. I believe the events recorded in Genesis chapter 7 provide just such a framework. On the first day of the global flood of Genesis (Genesis 7:11), the Scripture tells us that the "fountains of the great deep burst open." The word translated "burst open" is a violent word meaning to rip open or to tear apart.

In other words, it just didn't rain for 40 days and nights. The earth experienced geologic upheavals the likes that had never before occurred! The phrase implies that huge gashes or rips tore open the earth and that as a result large amounts of molten material and hot water erupted with incredible pressure and violence. Looking at it that way, it is not hard to see that calderas, these monstrous volcanoes of the past, could have been produced as a result of this one significant geologic event. No wonder geologists have called them *super volcanoes.*

The rim of the Yellowstone crater can be clearly seen in the background; in a caldera explosion, the main crater of the caldera apparently collapses and fills with rhyolite lava.

In a caldera explosion, after an explosive release of pyroclastic material in the form of ash and tuff, the volcano apparently collapses. What follows are successive lava flows predominantly of rhyolite lava.

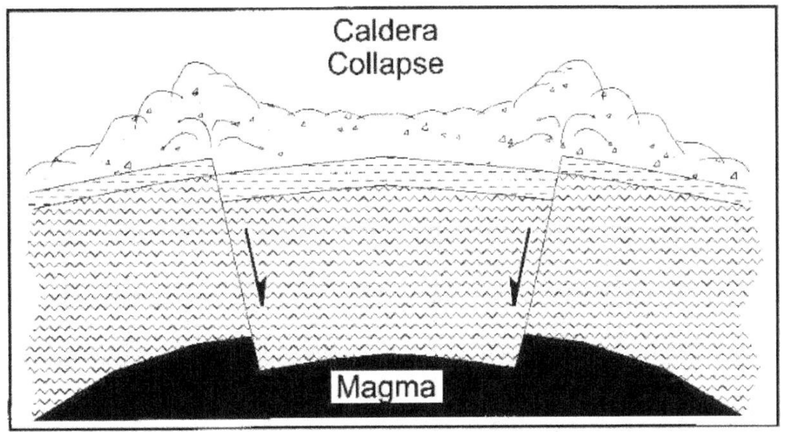

Figure 1. The ring fracture model for the eruption of ignimbrites and subsequent caldera collapse (after Cas and Wright, 1987).

The lava primarily erupted from the Yellowstone eruptions was a type of lava called rhyolite. It is lava high in quartz and therefore considered to be very explosive in nature. Quartz causes the lava to be highly viscous, and consequently, creates more explosive eruptions. Viscosity is a measure of resistance to flow. So, lava that is high in viscosity flows very slowly.

Rhyolite lava high in quartz (silica) and therefore extremely viscous; in other words, it flows very slowly, often preserving flow patterns of the lava, as you see in the photo; the term *rhyolite* is from the Greek, *to flow*. This picture was taken near the Grand Canyon of the Yellowstone.

This lava is not like the lavas being commonly extruded today. The most prominent kind of lavas erupted today are basalt and andesite, typical of Hawaii and the Cascade Volcanoes. Basalt is low in quartz and high in the darker minerals. It therefore flows with little viscosity as opposed to exploding.

Black basalt lava characterizes the active lava flows and eruptions on the big island of Hawaii.

This picture is of Mt. Shasta and Shastina in California, two andesitic, active volcanoes in the Cascade Mountains. These volcanoes build high cones of repeated andesitic eruptions and lava flows. Andesitic volcanoes although explosive, are not as explosive as rhyolitic volcanoes, because they contain less quartz. Andesitic volcanoes contain more quartz than basaltic volcanoes.

The number one question that I get asked on my Yellowstone field trips is, "Will Yellowstone erupt again and if so, when?" I find the most recent Yellowstone Resources and Issues Handbook to be rather interesting. In spite of the hype and Hollywood movies produced about Yellowstone over the last ten years, the official Park Handbook states, "Another caldera-forming eruption is theoretically possible, but it is very unlikely in the next thousand or even 10,000 years. Scientists have also found no indication of an imminent smaller eruption of lava."[2]

Although no one knows the future of Yellowstone, it does seem to be related to a past *unique* geologic event and therefore a huge eruption would not be likely in the near future, if ever again. The Genesis Flood is a reasonable and satisfactory explanation for this unique historical eruption of a giant volcano – a Super Volcano.

The continuing effects from the past eruptions of Yellowstone

Geologists consider Yellowstone to be an active volcano. That means that there are ongoing geologic disturbances related to this volcano.

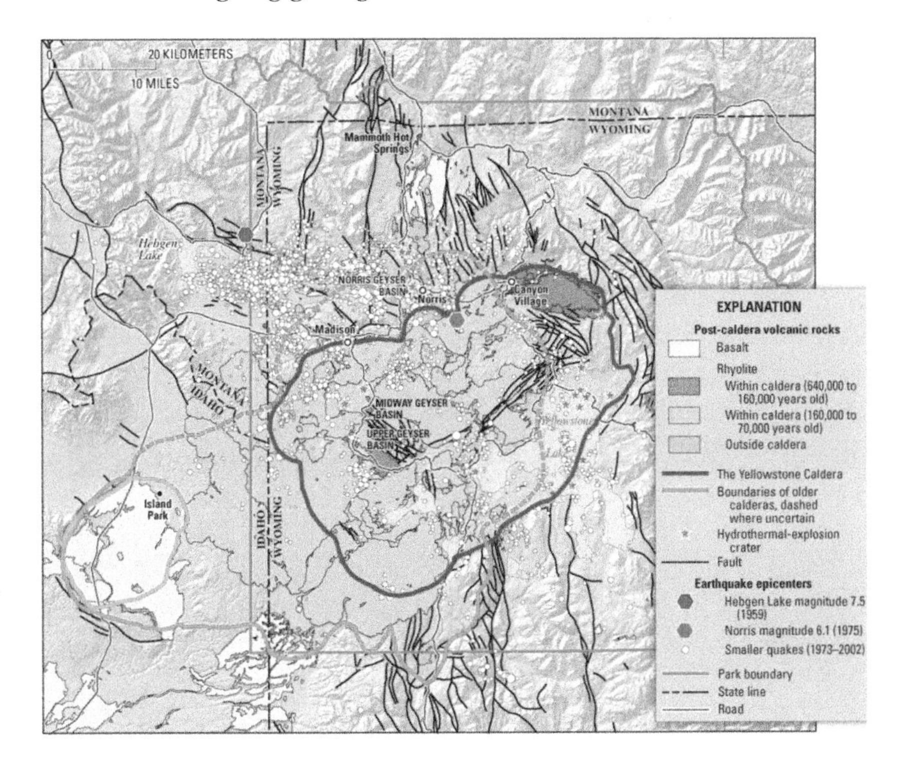

These include:

⇨ Earthquakes and faults – lots of them! Above is pictured a map showing the number of known earthquakes (red, pink and white dots)

[2] Yellowstone Resources and Issues Handbook, 2014, p. 102.

and faults (long black lines). Faults are fractures in the earth/rock where movement occurs between the two sides of the fracture. As you can see from the map, earthquakes are a normal part of Yellowstone National Park – as many as nine earthquakes a day! Many of these are insignificant, but some have been very destructive altering the thermal features in the Park as well as rearranging the landscape, similar to the 1959 Hebgen Lake earthquake. You can read more about this in Chapter 9.

- Ground swells or deformations called resurgent domes – geologists are not totally sure whether these ground swells are due to the movement of magma underneath the earth or to the ebb and flow of the underground thermal activity.

Sour Creek Resurgent Dome in the background has been continuously monitored for several years. The dome has undergone swelling and deflating.

- Active thermal features including geyser eruptions, hot springs, mud pots and fumaroles. Yellowstone continues to be a dynamic geologic area with thermal features continuously changing in intensity and frequency. One example is the West Thumb Geyser Basin. Once a very active geyser basin, it is now a steaming collection of hot pools and fumaroles with a few active hot springs. All these features will be explained in the next chapter.

Ever-changing West Thumb Geyser Basin

Chapter Five
The Thermal Features of Yellowstone

There are four different types of thermal features of Yellowstone. These include **geysers**, **hot springs**, **fumaroles** and **mud pots**.

Geysers – *Geyser* is an Icelandic word meaning, *to gush.* A geyser is essentially a hydrothermal (hot water) explosion. Most geysers are located in the proximity of active volcanic areas. It is believed that water seeps down through somewhat porous rock, comes into contact with hot rock that has been heated by its proximity to magma, builds up pressure and is therefore forced through a crack or crevice in the overlying rock and out to the surface.

Riverside Geyser, Upper Geyser Basin, Yellowstone Park

There are cold-water geysers too. Cold-water geysers' eruptions are similar to that of their hot-water counterparts, except that CO_2 bubbles drive the eruption instead of steam. An example of a cold-water geyser is Crystal Geyser in Utah near Green River, Utah.

Crystal Geyser, Green River, Utah

Over one thousand known active geysers exist worldwide. A study that was completed in 2011 found that at least 1,283 geysers have erupted in Yellowstone National Park, and an average of 465 geysers are active there in a given year. They are the main reason that Yellowstone was established as a protected national park in 1872. According to the Encyclopedia Britannica, approximately half the world's total number of geysers are in Yellowstone, about 200 on the Kamchatka Peninsula in the Russian Far East, about 40 in New Zealand, 16 in Iceland, and 50 scattered throughout the world in many other volcanic areas.

How does a geyser work?

The official Park literature defines a geyser as a *hot spring with constrictions.* Although it is not exactly known how a geyser works, geologists suspect that a geyser erupts when pressure, steam and a large supply of hot water combine under just the right conditions to produce a jet-like eruption.

Further the Park states that an eruption ceases when the water reservoir is depleted or when the system cools.

The main components needed for a geyser to work seem to be:

- ⇨ A continuous supply of water
- ⇨ An underground source of heat thought to be related to hot magma in the magma chamber of the Yellowstone Caldera
- ⇨ Pressure and carbon dioxide combined with steam
- ⇨ A plumbing system: constricted passageways through which superheated water passes on its way to the surface. The greater the constrictions, the more pressure is built up resulting in sometimes-magnificent eruptions of hot steam and water.

Beehive Geyser, located in the Upper Geyser Basin, near Old Faithful: if one is fortunate enough to be at the right place at the right time, one can see a magnificent eruption anywhere from 150 to 200 feet in height and lasting for about five minutes. Beehive Geyser is unpredictable and can sometimes go for weeks before erupting.

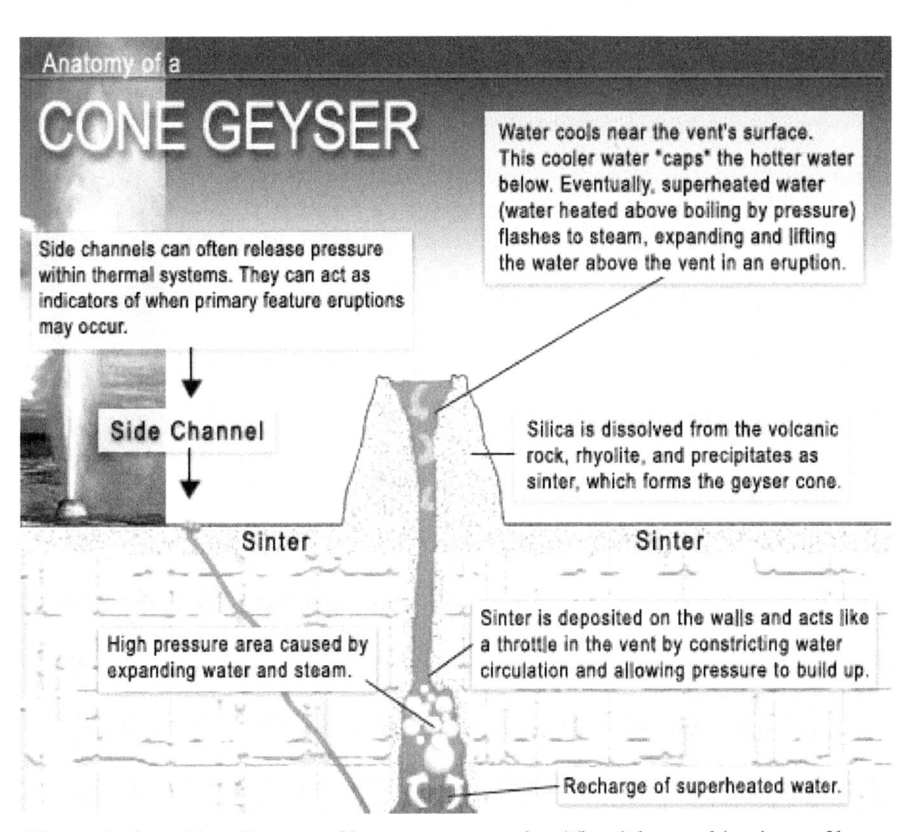

Theoretical working diagram of how a geyser works. The right combinations of heat, water source, pressure and plumbing system seem to be the ticket for a geyser eruption.

Remember that no one really knows what a plumbing system for a geyser looks like. It is far underground and is filled with hot water and steam. An experiment was conducted in 1992-1993 where a video probe equipped with temperature and pressure sensors was lowered into Old Faithful Geyser after an eruption had ended. At 22 feet, there was a narrow slot about four inches wide. Just below that was a wider area that contained water that was relatively cooler at 176°F. Then at 35 feet down, the probe reached an area about the size of a large car. As the geyser refilled with water, the temperature of the rising water was 244°F, 45°F hotter than the boiling point of water at that altitude. The probe was never lowered deeper than 46 feet into Old Faithful, so we really don't know what lies below the very top portion of a geyser. One other thing that the experiment showed was that much of the plumbing system was coated with a water and pressure-tight lining of geyserite. Geyserite is siliceous sinter derived from the rhyolite rock that is also composed of silica.

Siliceous sinter or geyserite is the white to gray crust found around and in the thermal features of Yellowstone. It is derived from the rhyolite lava rocks below ground where hot somewhat acidic water dissolves the rhyolite, carries it through the plumbing system and on to the surface. On its journey the siliceous sinter is precipitated out as the water cools. Precipitation is a process that occurs as a combination of gravity, temperature change and chemical reaction.

The geyserite or siliceous sinter cone of Castle Geyser built up over repeated eruptions of silica-rich water

Where does the water come from to continually fuel geyser eruptions? It used to be thought that geysers were continually fed by melting ice, snow and rain. Whatever these contributions are, they do not fully account for the continuous and consistent eruptive cycles of the geysers in Yellowstone where hundreds of millions of gallons of water are erupted each year. Many of the geysers of Yellowstone, such as Old Faithful, run just about like clockwork regardless of drought or wet seasons. In fact, the factors that seem to affect a geyser's eruptive cycles do not seem to have anything to do with the supply of water, but rather with earthquakes

that can disrupt the plumbing system of the geyser and mineral build-up within the walls of the plumbing system. It is now thought that perhaps there is some kind of a large aquifer below the earth. The amount of water needed to fuel the more than 10,000 thermal features in Yellowstone, especially the hot springs and geysers, is enormous!

The question I have pondered in relation to geysers is what part the Genesis Flood might have played in the underground source of water. The Biblical chronology of the Genesis Flood makes it a fairly recent event in history – about 4,500 years ago. Toward the end of the Flood, according to Psalm 104:5-9, the waters were standing above the then existing mountains. At God's rebuke, the waters fled to the place He had designated for them; the mountains rose and the valleys sank down. Where did the water go that had covered the earth? They probably went into two places. **(1)** They receded into the sunken valleys, now the ocean basins, and, **(2)** they receded into the places from where they came – into what are now aquifers: from the breaking up of the fountains of the great deep.

Hot springs - This may surprise you, but there is no universally accepted definition of a hot spring. In fact, I have discovered over 12 different definitions of a hot spring. So, here, I will be concerned with the hot springs of Yellowstone.

Hot spring vs. geysers

What is the difference between a hot spring and a geyser in Yellowstone? Hot springs are the most common feature in Yellowstone. The main difference between a geyser and a hot spring seems to be in their plumbing system: the underground passageways through which superheated water travels to the surface. The geysers all seem to have constrictions that aid in the build-up of pressure which helps produce an eruption. The hot spring's constriction seems to be fairly free of constrictions. Because of this, as the hot water reaches the surface, it cools and sinks, and then is replaced by hotter water from below. This process of circulation is called convection. This explains why the hot spring is always hot. Unless earthquakes or some other activity that produces constrictions changes a hot spring's plumbing system, there will not be eruptions, only continual flow of hot water sometimes spilling over into streams of hot water running out of the hot spring.

Some hot springs are pools. They look like puddles but are indeed active and very hot.

The most famous of all the hot springs in Yellowstone is Grand Prismatic Spring, famous for its vivid and beautiful colors produced by the various microbes that thrive in the hot wet environment.

Fumaroles – The word fumarole comes from a Latin word, *fumus*, meaning smoke. A fumarole is essentially a steam vent. The vent emits steam and gases such as carbon dioxide, sulfur dioxide, hydrogen chloride, and hydrogen sulfide. The steam is created when superheated water flashes to steam as its pressure drops when it emerges from the ground. This is due primarily to the lack of water available within the fumarole. There are an estimated 1,000 fumaroles within Yellowstone National Park boundaries. Fumaroles are considered to be the hottest thermal features in the Park.

Fumaroles at Norris Geyser Basin, Yellowstone

Roaring Mountain – the result of a monstrous hydrothermal explosion. It used to roar with the sounds of the hissing steam fumaroles. Now, one has to listen closely to hear the sounds of the various fumaroles. Such is the dynamic nature of Yellowstone.

Mud pots – Mud pots are extremely acidic hot springs with a pH (1.5) approaching that of battery acid. The primary cause of the mud formations of the mud pots is due to the presence of (1) a lack of water, (2) hydrogen sulfide gas, (3) microbes that convert the hydrogen sulfide gas into sulfuric acid. The sulfuric acid breaks down the geyserite, produced by the hot spring, into clay. The combination of these ingredients produces a thick mud which bubbles, as carbon dioxide is released into the muddy mixture. The acidic thermal features of Yellowstone are well-known for their rotten egg smell because of the hydrogen sulfide gas.

Mud pots are a delightful feature at Yellowstone. Anticipating the next carbon dioxide release in the form of bubbling mud is a fun experience at these hot springs.

Mud Volcano located just a few miles south of The Grand Canyon of the Yellowstone.

Chapter Six
The Microbes of Yellowstone

What are microbes? Microbes are single-cell organisms so tiny that millions can fit into the eye of a needle. They are living things. Microbes are everywhere. There are more of them on a person's hand than there are people on the entire planet! Microbes are in the air we breathe, the ground we walk on, the food we eat—they're even inside us! We couldn't digest food without them—animals couldn't, either. Without microbes, plants couldn't grow, garbage wouldn't decay and there would be a lot less oxygen to breathe. They seem an indispensable part of our planet – evidence of the design of a caring God! Microbes are extremely intricate creatures. They have DNA. They have complex parts that coordinate together. At one time, evolutionary biologists thought these creatures were *simple.* Not anymore! They all exhibit the characteristic of a system. A system is a set of interacting or interdependent components forming an integrated whole. Either these creatures were working as they do from their beginning or not at all!

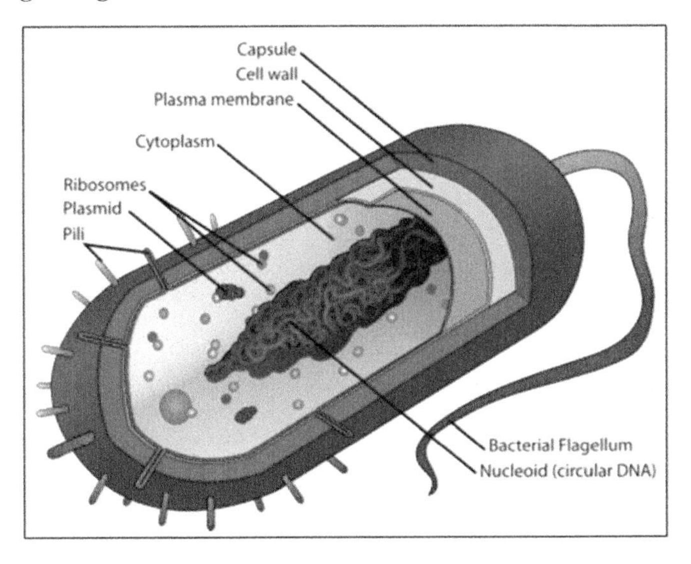

A typical prokaryote (bacterial) cell

Instead of looking at these amazing creatures as a *kind* created by God, biologists link them all together into an evolutionary tree of life.

Phylogenetic Tree of Life

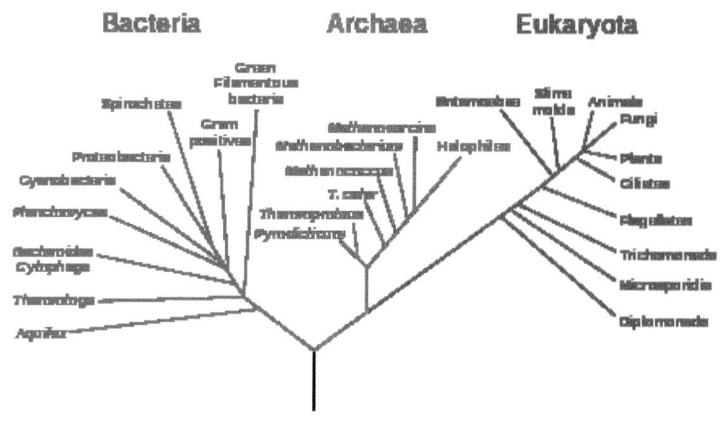

In the proposed tree of life for, evolutionary biologists classify all living things as sharing a common ancestry. In other words, all living things are related. The solid lines, by the way, are hypothetical. They have never been proven to be true. The impression is nevertheless left that all living things are related and therefore linked to a common ancestor.

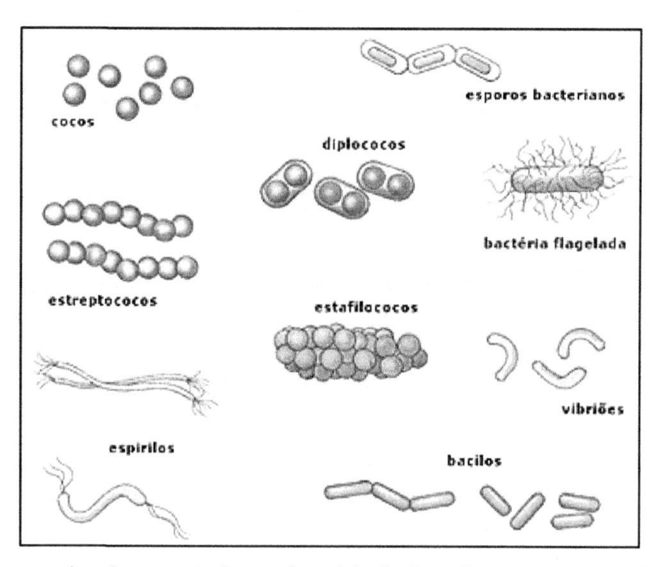

The various microbes are independent kinds that share a common Designer. Common ancestry is simply a belief. There are unique kinds with lots of variation within their kinds.

What is the difference among the three groups on the hypothetical phylogenetic tree of life according to modern biologists?

1. Prokaryotes - A prokaryote is a single-celled organism that lacks a membrane-bound nucleus, mitochondria, or any other membrane-bound organelle. The word prokaryote comes from the Greek *pro- (before)* and *karyon* (*nut* or *kernel.*) The Prokaryotes are then divided into two groups the **bacteria** and the **archaea**. Bacteria are sometimes called Eubacteria and archaea are sometimes called archaebacteria. The bacteria and archaea collectively are referred to as prokaryotes.

2. **Eukaryotes –** A eukaryote is any organism whose cells contain a nucleus and other structures (organelles) enclosed within membranes. The defining membrane-bound structure that sets eukaryotic cells apart from prokaryotic cells is the nucleus, enclosed by the nuclear envelope, which contains the genetic material. Since man has cells with nuclei, he is considered to be a eukaryote.

Classifying these critters falls into a branch of biology called taxonomy. The distinction of the title of the father of modern taxonomy has been given to Carl Linnaeus (1707 – 1778). Linnaeus was a brilliant Swedish botanist, physician, and zoologist, who laid the foundations for the modern biological naming system of living things. In addition, he is also considered one of the fathers of modern ecology. He was thought to be a giant in taxonomy in his day. The Swiss philosopher Jean-Jacques Rousseau said of him, "I know no greater man on earth." The German writer Johann Wolfgang von Goethe wrote: "With the exception of Shakespeare and Spinoza, I know no one among the no longer living who has influenced me more strongly." Swedish author August Strindberg wrote: "Linnaeus was in reality a poet who happened to become a naturalist." Among other compliments, Linnaeus has been called *The Prince of Botanists, The Pliny of the North*, and *The Second Adam*: quite the distinctions.

As is true of most of the scientists during this period of time, Linnaeus believed in the Scriptures as inspired by God and therefore believed that the Book of Genesis was historical. Linnaeus, like most of his colleagues, believed that God's creation could be discovered and organized. He therefore attempted to organize life into categories based on shared physical characteristics. While his work was significant and must have been a daunting task to undertake, he only partially delved into the Genesis created kinds. And in fact, it is readily admitted by biologists today that Linnaeus' system has been brought up to date to be more in line with Darwinian ideas. Two of those modifications have been, **(1)** to view some life as extremely primitive and some life as very complex; a hierarchy of evolution and not created by God, and, **(2)** to group the archaea (a Greek word meaning *ancient*) as descendants of the first forms of life on earth. Linnaeus would never have subscribed to these modifications.

Carolus Linnaeus

The Microbes of Yellowstone

Did you know that the thermal features of Yellowstone are teeming with microbes? The study of microbes at Yellowstone began fairly recently, during the 1960s. Why would scientists be interested in studying microbes?

In the Yellowstone National Park approved book, Seen and Unseen, Discovering the Microbes of Yellowstone[3], the authors give some reasons for studying the microbes in Yellowstone. While medical benefits derived from this study are certainly part of the motivation, the reasons most elaborated on were to find out if life exists on other planets, and to find out how life might have arisen in the assumed extreme environments of the early earth.

This belief in the evolution of life from primitive to complex has moved us in the wrong direction. If God created all life and if the early earth was as Genesis describes (a sphere of water), then all of Darwinism has been a deception of the highest order. Are microbes the leftovers from the

---

[3] Kathy Shehan, David Patterson, Brett Hanson, Brett Dicks, Joan Henson *Seen and Unseen, Discovering the Microbes of Yellowstone* (Montana State University: 2005), p. 99-100.

earliest forms of life as biologists insist today or are they a complex form of life originally created by God to aid man's existence?

Understanding the Terminology

While biologists have known about algae for quite some time, the microscopic world of bacteria and archaebacteria is a fairly new field. Yellowstone is home to a diverse community of bacteria and archaebacteria. Of particular interest to biologists are those that live in the hot springs of Yellowstone. These are called thermophiles or *heat lovers*. These are able to live at temperatures of over 113° F. There are even those capable of living at temperatures approaching 176°F! These are called *extremophiles*. Together these thermophiles account for most of the beautiful array of colorful patterns seen in the various hot springs throughout Yellowstone.

The most celebrated laboratory for the study of thermophiles is Grand Prismatic Spring. Each color represents a community of amazingly complex living organisms.

Grand Prismatic Spring

Grand Prismatic Spring

Microbes in Chromatic Pool in the Upper Geyser Basin

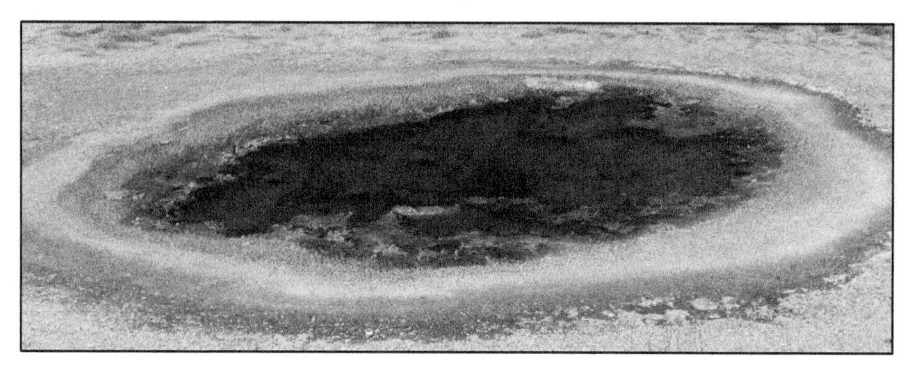

Microbes in a hot spring in the Upper Geyser Basin

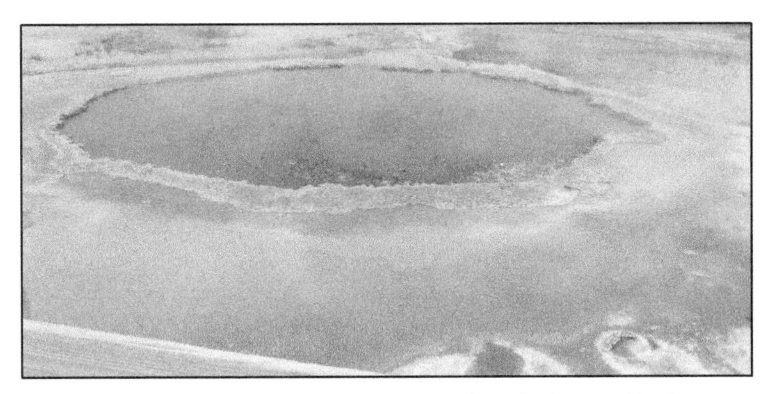

Microbes in a hot spring at West Thumb Geyser Basin

Microbes in a hot spring in West Thumb Geyser Basin

Microbes in West Thumb Geyser Basin

Microbes at Mammoth Hot Springs, Yellowstone

Microbes at West Thumb Geyser Basin

Microbes at the Firehole River, Midway Geyser Basin

The pH scale and Microbes

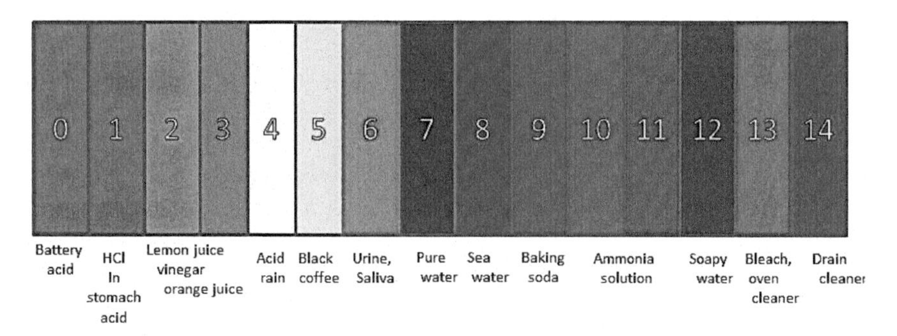

The pH Scale
0-6 is acid and 8-14 is base, with 7 being neutral, in between the two.

The pH scale is a device used to measure the acidity or alkalinity of a substance. Acids and bases (alkali) are extremely important to the governing and maintenance of living things. If too much of an acid or base is consumed by living things, it will be destructive. Notice that the great *equalizer* in the pH scale is water! It is my personal belief that this is no accident. It speaks of a design that helps keep our world in balance. And so far as we know, the earth is the only place with the abundance of water that we have.

The pH scale is used by biologists to identify the various microbes in the Yellowstone thermal features. The different thermal features of Yellowstone are divided into those that are acidic, meaning they have a low pH and those that are basic or those that have a higher pH.

The microbes that live in hot acidic features are called *thermoacidophiles.* They not only love the heat, they love the acidic conditions! Many of these thermoacidophiles live in conditions with a pH of around 1!

Black Dragon's Caldron: this area has some of the most acidic and dynamically changing features in the Park. The pH approaches that of battery acid!

The thermal features of Norris Geyser Basin are extremely acidic. The whole basin wreaks of rotten egg smell which is a combination of the interaction of the thermoacidophiles and the sulfuric acid.

Chapter Seven
The Rocks and Minerals of Yellowstone

And now we come to probably what is the most interesting to me – the rocks and minerals of Yellowstone. The Greater Yellowstone region has more diversity of the rock types than just about any one area of the world. And the region to the northwest of Yellowstone Park has some of the most significant limestone formations anywhere.

Limestone, Big Sky, Montana

The Gallatin and Absaroka Mountain Ranges, which immediately surround the north and northeast of Yellowstone Park contain the most petrified logs in volcanic ash flows in the world.

Electric Peak in the Gallatin Mountain Range, above Gardiner, Montana

The Gallatin Mountain Range contains an almost infinite number of petrified logs.

Petrified logs from the Gallatin Mountain Range, northwest of Yellowstone

The Beartooth Mountains to the north of Yellowstone are some of the most studied mountains in the world. They are filled with plutonic and metamorphic rocks, and show extreme uplift in the past.

The majestic Beartooth Mountains in Montana full of granite, a plutonic rock, and gneiss, a metamorphic rock

The Absaroka Mountains wrap around the park from the north to the southeast and are estimated to contain upwards of 9,200 cubic miles of volcanic lava, ash and tuff – the largest of any volcanic activity in the world!

Pilot Peak and Index Peak: These are part of the eroded remnants of the Absaroka Mountain Range; some of the greatest pyroclastic flows on record – 9,200 cubic miles of volcanic debris!

The volatile Madison Range with its history of earthquakes and metamorphic rocks

The Rock Types

Here is just a brief review for those who are trying to figure out the names and composition of the various rock types. Secular geologists have divided the rock types into groups by how they *think* the rocks formed. But I don't believe that is an acceptable way in light of the Scriptural view of earth history. The primary reason is that the history of most of the rocks was a thing of the past – the distant past for secular geology. No one saw the various rock formations form. No one saw the crust of the earth form. No one saw the metamorphic rocks form. No one saw the granites form. Their formation is interpreted by ideas. The prevailing idea in secular geology is a 4.6-billion-year-old earth. The idea from Genesis is a young earth of no more than 6,000 years old. The framework we will use in this guide will be that of Genesis chapters 1-8.

Secular geologists have traditionally divided the rock types into three groups:

1. **Igneous Rocks –** rocks geologists think were formed by fire or heat.
2. **Metamorphic Rocks** – rocks geologists think were formed over millions of years by heat and pressure deep underground.
3. **Sedimentary Rocks** – rocks geologists think were formed over millions of years of deposition, erosion and transportation of sediments.

The only rocks observed to be forming by geologists have been the volcanic rocks. What about granite? What about gneiss and schist? What about the huge limestone formations found around the world? Nope – no one has seen these rocks being formed.

The White Cliffs of Dover: a large limestone chalk formation in southern England

Let's take a second look at these groups in light of the young earth and global flood presented in the Book of Genesis. We will divide the rock types into the following groups:

1. **Plutonic Rocks** – rocks that were formed initially as the foundation or basement rocks of the earth.

2. **Volcanic Rocks** – rocks that have been and are forming as the result of volcanic activity. These rocks have all been formed during and after the Flood. Volcanic rocks continue to be formed as remnants of the great Genesis Flood.

3. **Metamorphic Rocks** – the word metamorphic means *change*. These rocks appear to have been changed by some process that has stretched and rearranged the minerals into different kinds of rocks. While these could have been part of the foundation rocks created by God at the creation of the earth in Genesis chapter one, they could also be the product of the tectonic activity associated with the Flood.

4. **Sedimentary Rocks** – rocks that have been laid down in/with/by limy mud and watery sand and clay sediments. The keys here are *water* and *mud.*

Yellowstone Park and the Greater Yellowstone Ecosystem have all of the rock types!

The Rocks of the Yellowstone Area

Rocks are made of minerals. Typically, there are six light and six dark minerals that comprise most of the rocks in the Yellowstone area.

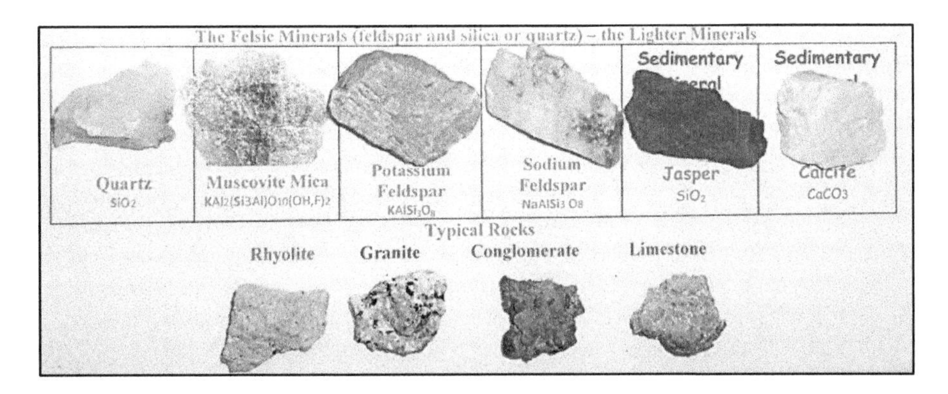

Granite – Granite is a plutonic rock. The word plutonic comes from the Roman god of the underworld, *Pluto***.** So, these rocks are called *basement* rocks by secular geologists. These form the continental crust of the earth. Secular geologists also add the metamorphic rocks to this, but their formation has not been observed. Considering the Flood framework, it might be that plutonic rocks were deformed by the tectonic forces of the Flood; the plutonic rocks changed into metamorphic rocks and mixed with the plutonic rocks. This certainly appears to be the case in the Beartooth Mountains. Secular geologists date these rocks at between 2 and 3 billion years old by way of radiometric dating. They are made largely of quartz, biotite or muscovite mica and potassium feldspar. They most often appear as a gray to pinkish mass from the potassium feldspar.

Quartz, biotite mica, muscovite mica and potassium feldspar

Granite exhibits an even mix of potassium feldspar (either pinkish or white), quartz and either biotite or muscovite mica. It is considered to be a light-colored plutonic rock.

Gneiss – Gneiss is a metamorphic rock common in the Beartooth and Madison Range of Mountains. It is typically banded (striped) with alternating bands of quartz, mica and feldspar.

Quartz, biotite mica, muscovite mica, feldspar

Gneiss from the Madison Valley

Gneiss from the Gallatin Canyon along US Highway 191 toward West Yellowstone

Amphibolite – occurring along with the plutonic rocks in the Beartooth Mountains is a metamorphic rock called amphibolite. As the name implies it is predominantly made of amphibole or hornblende. It is a dark rock displaying the classic hornblende crystals.

Amphibole

Amphibolite

Rhyolite - As Yellowstone National Park is mostly contained within a caldera, most of the rocks are going to be volcanic in makeup. The most abundant volcanic rocks in Yellowstone are **rhyolite, rhyolite derived rock** and **rhyolite tuff**.

The word rhyolite comes from a Greek word meaning *flow.* The composition of rhyolite is mostly quartz and potassium feldspar. Because of its high concentration of quartz, (upwards of 70%), rhyolite lava will flow with a high degree of viscosity. Viscosity is a word that describes the degree of resistance to flow. Hence the word, rhyolite or *flow.* The lava will generally leave flow patterns in the lava as it cools. The lava typically does not move far from its cone or fissure. And typically the rhyolite eruptions are extremely explosive.

Rhyolite mostly consists of the following rock-forming minerals:

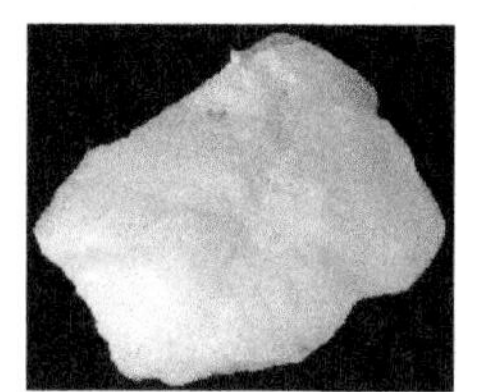

Quartz, biotite mica, muscovite mica, and potassium feldspar

Rhyolite from the Grand Canyon of the Yellowstone

Samples of rhyolite lava from West Yellowstone, Montana

One of the most interesting places to visit is the Grand Canyon of the Yellowstone. The Canyon is up to 1,200 feet deep and up to 4,000 feet wide. As the Yellowstone River flows north from Yellowstone Lake, it leaves the Hayden Valley and plunges first over Upper Yellowstone Falls and then a quarter mile downstream over Lower Yellowstone Falls, at which point it then enters the Grand Canyon of the Yellowstone. Both of these falls are breath-taking!

Upper Falls (109 feet); and Lower Falls (308 feet) of the Yellowstone

But the sight that catches my attention, as a geologist, are the Canyon walls. The colors are striking. The Grand Canyon of the Yellowstone consists of rhyolite ash and lava that has been altered. Altered by what? Chemical reactions that involve volcanic rock, iron oxides, hot water and steam create the bright yellow, orange, and red colors that are visible in the walls of the Grand Canyon of the Yellowstone. This process is called hydrothermal alteration.

The Canyon Walls, cut away by great quantities of rushing water, have exposed what really went on in the past history of Yellowstone. Immense deposits of rhyolite ash and lava were laid down in one of the largest volcanic eruptions in earth history. Some people think that the Yellowstone got its name from the light colored hydrothermally altered rhyolite, but that is not certain.

Trachyte – Trachyte is potassium-rich lava, common in Yellowstone and the surrounding region. It is almost exclusively potassium feldspar crystals. The crystals are displayed as blocky light-colored crystals in a fine-grained texture.

Potassium feldspar

Trachyte lava – found in the Beartooth Mountains north of the Absaroka Mountains that immediately surround Yellowstone.

Obsidian – Obsidian is volcanic glass: pure quartz colored by iron. Obsidian is actually a type of rhyolite lava, high in quartz and therefore glassy in appearance and texture.

Although Obsidian is usually black in appearance, it is made from pure quartz but it has been colored by iron.

Quartz

Obsidian with flow bands from Obsidian Cliff, Yellowstone, just a few miles south of Mammoth Hot Springs. The yellow-green color is the lichen that grows on the rocks in Yellowstone.

Obsidian is most often a black volcanic glass; technically it is rhyolite lava colored by iron.

Much of the obsidian in Yellowstone comes from Obsidian Cliff. It is located just eight miles north of Norris Geyser Basin and is well worth the stop. Obsidian cliff is made of five square miles of volcanic glass. Native Americans from all over the United States came to mine this volcanic glass for the arrowheads and spear points.

Obsidian cliff – it is hard to imagine that this black shiny volcanic glass formation was once a moving hot lava flow!

Obsidian boulder from Obsidian Cliff, Yellowstone

Obsidian flow in the form of columnar jointing at Obsidian Cliff. Jointing (the formation of natural cracks) seems to occur when lava begins to cool and contract along joints. It is an interesting phenomenon and one that is not totally understood.

Other areas around the Park contain obsidian boulders and sand. The areas around Old Faithful and West Yellowstone are particularly covered in black glassy sand. Lake Yellowstone exhibits this sand on its beaches.

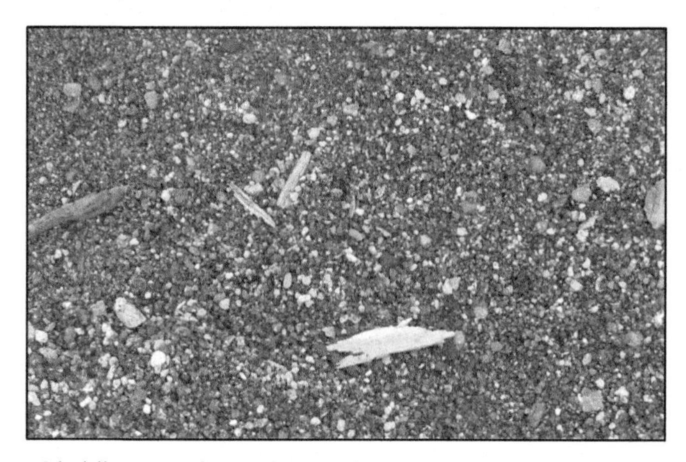

Obsidian "sand" on the beach of Lake Yellowstone near West Thumb Geyser Basin

Vitrophyre – Vitrophyre is another volcanic rock. It is a form of obsidian, which has undergone hydrothermal alteration (contact with water while still hot). It is characterized by brittleness and large visible phenocrysts. This is especially concentrated around West Yellowstone where combination of rapid glacial melting and rhyolite eruption must have been profuse. The entire valley west of Yellowstone is covered in this obsidian.

Vitrophyre from California; Vitrophyre close-up, near The Grand Canyon of the Yellowstone

Tuff – Tuff is a type of volcanic rock that consists of consolidated ash welded together, often into a very hard rock. It is sometimes referred to as welded tuff. In a volcanic eruption, hot steam, volcanic gasses, rock particles (called tephra) and lava are ejected far up into the atmosphere. The coarser material falls first and then the finer material. Because of the heat, these materials are welded together into a compact mass of a rock that looks kind of like oatmeal with nuts and brown sugar.

Typical pyroclastic tuff in Yellowstone Park

The white lenses (elliptical-shaped spheres) in this tuff are feldspar minerals. The black minerals are hornblende and pyroxene. All this material is bedded into solidified or welded ash.

Tower Fall: the spires (hoodoos) are eroded remnants of tuff

Huckleberry Ridge Tuff – Huckleberry Ridge Tuff is made up of bits and pieces of quartz, volcanic rocks, olivine, hornblende and pyroxene.

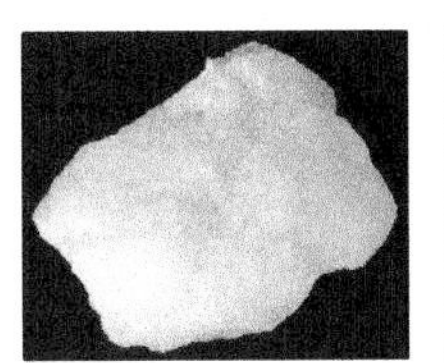  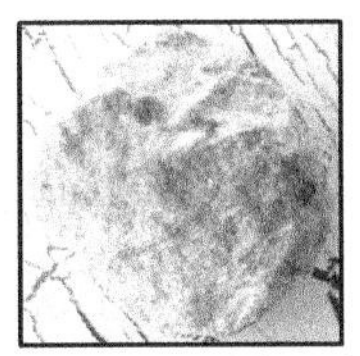 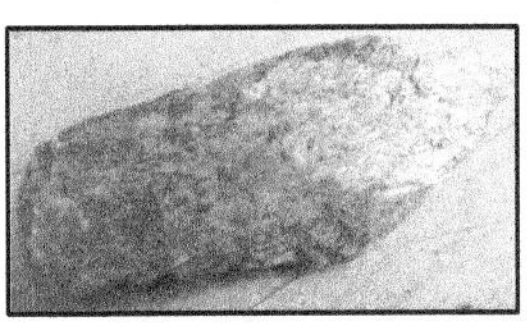

Quartz, pyroxene, olivine and hornblende (amphibole)

Rhyolite tuff at Sulphur Caldron

Rhyolite tuff

Huckleberry Ridge Tuff at the Golden Gate Bridge just south of Mammoth Hot Springs, Yellowstone; the whole wall is volcanic tuff. This is a massive amount of pyroclastic material!

Mesa Falls Tuff in Idaho, part of the Yellowstone eruptions

Lava Creek Tuff – located near Madison Junction, Yellowstone Park

Basalt – the word means *hard.* It is a very hard rock. It occurs in flows and is not associated with explosive volcanoes. Although not significant in Yellowstone, it is nevertheless present. It is generally dark gray or even black lava (from the dark rock-forming minerals) containing a significant amount of iron (magnetite), magnesium (olivine), calcium feldspar and pyroxene. In just the right conditions, basalt will oxidize and thus have the appearance of rust.

 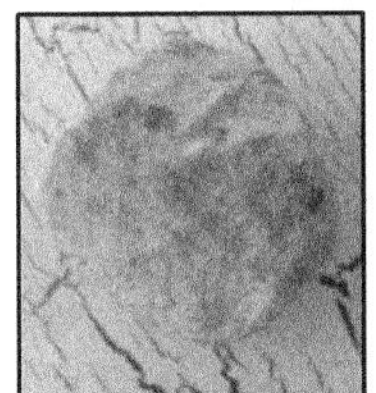 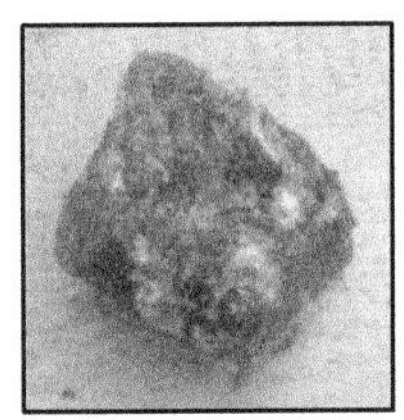 

Iron (magnetite), magnesium (olivine), calcium feldspar and pyroxene

Basalt at Tower Fall

Another form of basalt that occurs in Yellowstone is in columns, called columnar basalt. During a lava flow as the basalt lava cools, it forms fractures or contractional joints. This jointing often forms in regular hexagonal or pentagonal shapes. Good examples of this phenomenon can be found at Sheepeater Cliff and at Tower Junction.

Basalt columns – Sheepeater Cliff

Basalt tuff – As rhyolite eruptions produce tuff, so do basalt eruptions. These are most often characteristic of basalt flows, however, where bits and pieces of volcanic rock are incorporated into a lava flow.

This basalt tuff is made of bits and pieces of volcanic rock within a dark ash. This type of rock is most often referred to as volcanic breccia.

Volcanic breccia and conglomerate – Mud and basalt pieces were carried over many square miles in the mudflows of Yellowstone. Some of these flows show a tremendous amount of velocity, as many of the stones are rounded. Others are angular. The angular mix is called breccia. The rounded mix is called conglomerate.

Volcanic mud flow of breccia and conglomerate all along the road near Dunraven Pass, Yellowstone

Close-up of volcanic mud flow of breccia and conglomerate all along the road near Dunraven Pass, Yellowstone

Geyserite – Some consider geyserite to be volcanic, some consider it to be sedimentary, and others consider it to be a mineral. Geyserite is a product of silica precipitation derived from superheated water that contains carbonic acid. Precipitation is a chemical process where the silica is released and built up as a very hard rock on the surface. As the water rises through rhyolite lava rock, it dissolves the silica-rich rock in superheated water, carrying it to the surface where the silica precipitates out as siliceous sinter or geyserite. Geyserite is technically an opaline silica. That means that it is hydrothermally altered silica. Fresh sinter is white. In time, it changes to a drab gray color.

Geyserite or sinter as a precipitate at West Thumb Geyser Basin

Travertine – Travertine is a sedimentary rock. Travertine is a product of limestone precipitation derived from super-heated water that contains carbonic acid. Precipitation of travertine works the same as geyserite except that the hot acidic water rises through limestone instead of rhyolite. The thermal features at Mammoth Hot Springs just inside the North entrance to Yellowstone are the place to see the travertine deposits. Why is limestone involved? Limestone is a sedimentary rock, right? But Yellowstone is mostly volcanic.

The explanation for this is simple within a Genesis Flood framework.

The lime, sand and other soils and clays would have been torn up, transported and deposited within the first 150 days of the Genesis Flood. Yellowstone apparently erupted through this freshly Flood-laid and formed limestone. Mammoth is outside the main caldera, where so much of the limestone has been preserved. So as you move to the north within the caldera, the rock formations become limestone.

Mammoth Hot Springs

Travertine – a limestone rock that forms the Mammoth Hot Springs Terraces

A panoramic view of the Mammoth area; all of the rock you see in this picture, including the mountain ridge in the background is travertine that was deposited from hot spring activity at some time in the past. The current active terraces are but the remnants of a volcanic history that had been enormous shortly after Yellowstone erupted.

Petrified logs – Some consider petrified wood to be sedimentary rock because silicate minerals have replaced the original cells of the wood. The silica would have come from mineral-rich waters, produced by the significant amounts of volcanic ash from Yellowstone. Ash is essentially tiny particles of volcanic rocks and glass (silica). Logs would have been buried in this mineral-rich sediment, consequently the sedimentary designation. When people think of petrified logs or petrified trees, they usually think of the petrified *forests* located in the northeast part of Yellowstone – Specimen Ridge, part of the Absaroka Mountains. But these are only a small part of the enumerable quantities of petrified wood that is located in the Yellowstone region. Hundreds of thousands of petrified trees have been trapped in the volcanic mudflows in the Gallatin and Absaroka Mountains. All combined, there are more petrified trees in the Yellowstone region than anywhere in the world. It is huge.

Specimen Ridge, northeastern part of the Park

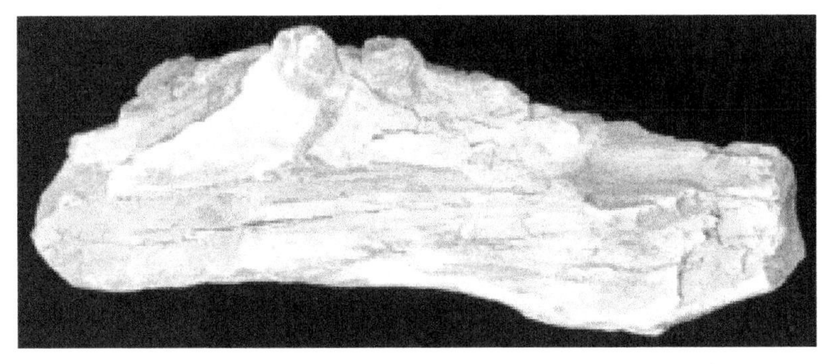

Petrified wood from the ash of the Absaroka Mountains

Limestone – Limestone is a sedimentary rock and it sits directly on top of the basement rock in Yellowstone and is covered by volcanic rock. This feature is exposed in many parts of the Park and is called by secular geologists, The Great Unconformity. Why is that?

The Great Unconformity
Absaroka Volcanic Rocks: 30-40 Million years old
Limestone: 200-300 Million years old
Basement Rock: 2-3 Billion years old

The basement rock is supposed to be 2-3 billion years old. The limestone that sits directly on it is supposed to be 200-300 million years old. The Absaroka volcanic rocks sit on top of the limestone, and are supposed to be 40 million years old. You do the math. There is a gap, a huge time gap, between the basement rock and the limestone and another significant time gap between the limestone and the volcanic rock. This represents over 2 billion years of geologic time missing! Where did it go?

Secular geologists explain that it either was never deposited, or it all eroded away. But the contacts between these gaps are flat, indicating that there was simply no erosion that took place. And how can you have this amount of time missing when you have the rest of it firmly in place? The secular explanation is simply a convenient way to account for the missing time or gaps.

Limestone from the Great Unconformity, Cooke City, Montana

The Great Unconformity outside of Cooke City, Montana, the Northeast Entrance of Yellowstone National Park. The Basement granite and gneiss is on the bottom. The exposed horizontal layer in the middle is the limestone and the Absaroka volcanics are on top. Notice that these rocks are separated by flat contacts. This would indicate that these rocks were laid down rather quickly and in quick succession.

Rocks of the Greater Yellowstone Region from the Author's Collection

There are a tremendous amount of interesting rocks that one can collect all around the Yellowstone region. My practice is to know where the boundaries are to the Park and then collect outside of those areas. I have found that this greatly reduces the temptation to collect within Park boundaries!

Please feel free to use these pictures as your own guide as you are touring around the Park. If you would like samples from the Yellowstone area, please contact me and I can arrange to get those to you.

Amygdaloidal basalt - Northeast Entrance; an amygdale is a cavity found particularly in volcanic rocks that is filled with a mineral crystal of some kind. (The white splotches are the mineral crystals.)

Aphanitic basalt - meaning that the individual mineral crystals cannot be seen with the naked eye; is a dark gray to black volcanic lava rock.

Lots of aphanitic basalt lava is found in the northern part of Yellowstone and throughout Gardiner, Montana

Basalt porphyry – is a type of lava that contains large visible mineral crystals of feldspar. This is called porphyry.

Basalt porphyry from the Northeast Entrance to Yellowstone

Andesite porphyry – Found at the Northeast Entrance; andesite porphyry is andesite lava with visible mineral crystals of usually sodium or andesine feldspar and/or pyroxene. Andesite can be anywhere from a light gray to dark gray to a brownish rock, easily identified by its larger phenocrysts (larger, visible crystals, usually feldspar) of feldspar and hornblende.

Andesite porphyry

Diabase – Found at the Beartooth Mountains – another name for diabase is microgabbro. It is a dark colored plutonic rock usually consisting of very small but visible phenocrysts. It is easily mistaken for basalt.

Diabase

Amphibolite – Found in The Beartooth Mountains – a metamorphic rock consisting of the rock-forming minerals, amphibole and plagioclase feldspar; schistose in appearance.

Amphibolite

Hornblende dacite, Wapiti Valley, out the East Entrance – Dacite is a light gray to dark gray volcanic rock consisting of quartz and plagioclase feldspar. Sometimes dacite exhibits the addition of hornblende (amphibole) crystals in its matrix. These are the black needle-shaped crystals in the picture below.

Hornblende Dacite

Tuff – Yellowstone Park is all about tuff! It is a volcanic pyroclastic rock originally in the form of very hot, swift-moving steam, glass and bits and pieces of broken volcanic rock. It sweeps along the ground moving at speeds of over a hundred miles an hour, stripping everything in its path. As it comes to rest and cools, it can form a very hard rock.

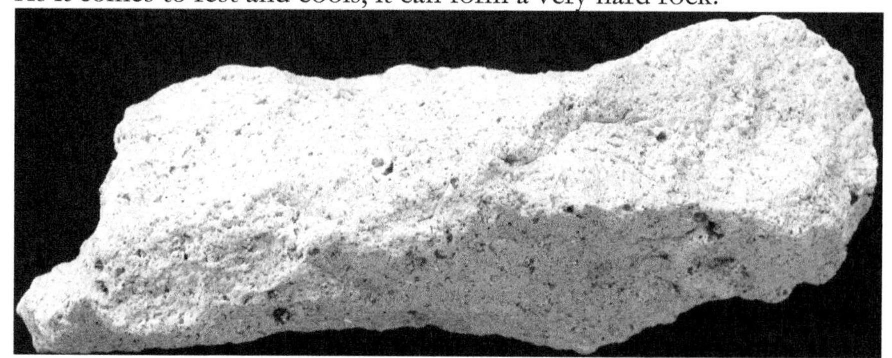

Tuff from the Northeast Entrance of Yellowstone Park

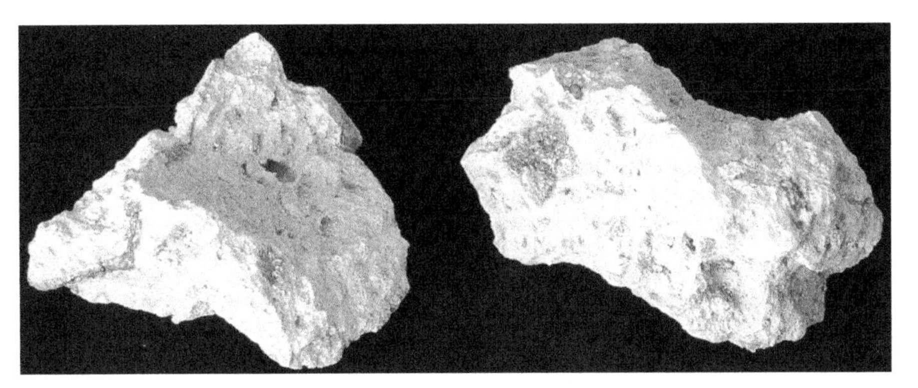

Tuff from the West Entrance of Yellowstone Park

Tuff from the Hebgen Lake area, Montana

Tuff from the Northeast Entrance to Yellowstone Park

Tuff from the South Entrance to Yellowstone Park

Rhyolite porphyry – rhyolite with visible feldspar phenocrysts

Samples of rhyolite from West Yellowstone, Montana

Obsidian, southern Montana – obsidian is a type of volcanic glass usually colored by iron. It consists primarily of quartz (glass). It is a type of rhyolite lava. No one really knows how it is formed.

Obsidian

Schist – biotite schist from the Beartooth Mountains. It is a metamorphic rock composed mostly of mica (either biotite or muscovite) in flakes. It has the appearance of glitter that has been glued onto a rock.

Schist

Vesicular basalt – basalt lava with an abundance of gas vesicles (gas pockets)

Vesicular Basalt

Dog tooth **calcite** – from Big Sky, Montana. Calcium carbonate in the shape of pointed crystals is embedded in a limestone wall. The limestone is from the Madison Limestone Formation covering hundreds of square miles of the Greater Yellowstone Ecosystem.

Dog-tooth Calcite

Chapter 8
The Ice Age in Yellowstone

What is or was an ice age? When people think of an ice age, they think cold – very cold. In fact, the picture that is conjured up is that of an icebox: so cold in fact that life disappears. Is this really what occurred in our brief earth history?

Secular geologists teach that not only was there AN ice age, but there were at least 24 of them spread out over millions of years, the most recent being that which geologists call the Pleistocene Epoch, covering from 2.5 million years ago, to about 11,700 years ago. They believe that it covered most of the northern hemisphere of earth, some of the southern hemisphere and the higher latitudes throughout the world. On the Geologic Time Scale, it would look like this:

Subdivisions of the Quaternary System

System/Period	Series/Epoch	Age in Millions of Years
Quaternary	Holocene	0.0117-present
	Pleistocene	Lasting about 2.5 million years

These subdivisions are then further divided into Stages and Ages. But that is a little more in depth than what we want to look at right now. The main period of time we want to explore now is the **Pleistocene Epoch**. That is what is most associated with the Ice Age.

The word Pleistocene means *more new* or *newer*. It is called that because secular geologists believe that there were other ice ages that preceded it. The ice age immediately preceding the Pleistocene was the Pliocene Epoch. What is confusing about this is the word Pliocene also means *more new* or *newer*. It is roughly translated, *continuation of the recent*. Charles Lyell originally coined the terms Pleistocene and Pliocene in 1839. The term Pleistocene was used to describe rock strata in Sicily that had fossils of mollusks in them that were just like those living today. He had originally thought that the Pliocene was the younger of the two divisions.

These have further undergone revisions since then by the International Union of Geological Sciences (IUGS). This organization is another interesting story. Obviously there has been a lot of exploration in geology since the days of Lyell. And this has spread to other countries. With all the diverse opinions arising, there was a need to codify geology. This is done by the IUGS. They make the official pronouncements that declare something to be the consensus of modern geology. In other words, we have added another layer of authority to modern geology! It is sounding more and more like a religious system all the time. OK – before we get lost in all the details, suffice it to say that there is plenty of geologic evidence to indicate that there was a period of time in earth history when there was a lot more ice and snow in our world than there is now.

An artist's impression of the global ice age maximum: there is abundant evidence that massive thick sheets once covered particularly the northern hemisphere and the higher latitudes of our earth.

Scandinavia exhibits some of the typical effects of ice age glaciation such as fjords and lakes.

Evidence of Past Glaciation

What is a glacier? A typical definition reads this way – *A glacier is a persistent body of dense ice that is constantly moving under its own weight; it forms where the accumulation of snow exceeds its melting. Glaciers slowly deform and flow due to stresses induced by their weight, creating crevasses, blocks of ice, and other distinguishing features.*[4] Present as well as past glaciation indicates that glaciers move and leave obvious signs of earth deformation as a result. These signs or evidence would include:

1. **Glacial moraines –** accumulation of unconsolidated soil and rock deposited as a glacier advances and retreats. It acts like a road grader.
2. **Glacial drumlins –** an Irish word meaning *littlest ridge*; it is an elongated hill in the shape of an inverted spoon or half-buried egg formed by glacial ice acting on underlying unconsolidated till or ground moraine.

[4] http://en.wikipedia.org/wiki/Glacier

3. **Glacial valley cutting** – a U-shaped valley produced by the movement of glacial ice.
4. **Glacial cirques** – a French word meaning *arena;* it is an amphitheater-like bowl formed from the erosive activity of moving glaciers. It looks like an amphitheater.
5. **Glacial till** – unsorted glacial sediment. Glacial till is the loose rocks, dust and boulders that have been moved by glaciers as they advance and retreat. (Glacial tillite is the same except that it has been cemented together into a unified hard rock).
6. **Glacial horns** – or pyramidal peak; an angular, sharply pointed mountain peak from the erosion due to multiple glaciers diverging from a central point.
7. **Glacial erratics** – comes from a Latin word, *errare*, meaning, *to roam or to ramble.* An erratic is a piece of rock or boulder that differs from the size and type of rock native to the area in which it rests. These show signs of having been transported from many miles by moving ice. They are generally not water tumbled (rounded), but angular.
8. **Glacial arêtes** – a French word meaning, *edge or ridge*; it is a thin ridge of rock that is left separating two valleys. It is produced when two glaciers on either side have created cirques, leaving the knife-edge ridge between.
9. **Past migration of certain animals and man** – typical ice age animals whose fossils have shown up in areas that are typically associated with warmer climates today.
10. **Glacial kettles** – a hole or pothole; it is a shallow, sediment-filled body of water formed by retreating glaciers or draining floodwaters.
11. **Glacial loess (pronounced three ways, LO-iss, less, or luss)** – originally a German word that was translated into English as loose. It is loose, fine, wind-blown sediment or dust that has accumulated into huge formations of loose sediment.

So, when we speak of the Ice Age, we are talking about a time in earth history when weather patterns were sufficiently different from that of today, leaving evidence of huge amounts of ice and snow that once occupied a considerable portion of the earth's surface. Determining the age and cause of that ice age is one of the biggest mysteries of modern geology. Many ideas have come and gone and yet no definitive scientific proof has been forthcoming. In fact, so cantankerous have the arguments been that the IUGS has had to officially state the age boundaries for this mysterious period of time in earth history.

A Biblical View

As modern science has not been able to solve this huge geological mystery, let's look at what the Bible might have to say about an ice age. First, the Bible does not mention an ice age. Many people have misunderstood this and therefore concluded that an ice age was just pure myth or a trick of science. Nothing could be further from the truth. Even though the Bible does not mention an ice age per se, it does give us clues as to how it might have come about. It does? Go back and read Genesis 7-8. In Genesis 7:11 we read that in the 600th year of Noah's life...all the fountains of the great deep burst open. This statement implies that the earth split into many sections. Out of these cracks, in what was then the ocean floor, must have come great amounts of lava. Further, great tectonic or earth movement must have been initiated by this geological upheaval. These events must then have precipitated profuse volcanic eruptions on land.

Now, something happens when great amounts of ash are spewed into our atmosphere. It blocks part of the sun's reflective energy and cools the atmosphere. The eruptions of Mt. Tambora and Krakatoa in the 1800s both produced so much ash that the temperature was lowered worldwide several °. As the magma/lava poured into the ocean and with the addition of hot water from the breaking up of the fountains of the great deep, the combination of increasing warmer water, increasing evaporation and a cooler atmosphere would eventually have produced a large amount of snow. As more snow began to fall and accumulate, ice would form, creating glaciers. It is one of those so-called perfect storms. More snow accumulated than melted. It is a perfect recipe for an ice age. The Genesis Flood would have produced all of these ingredients causing a unique and one-time ice age with glacial advances and retreats. As the volcanic action died down, the atmosphere would have cleared, allowing more of the sun's reflective energy to warm the earth and catastrophically melt the ice that had rapidly built up. Some creationists have proposed a period of about 200 years for rapid ice buildup and around 300 to 500 years for rapid, catastrophic ice melting.

So, what does all of this have to do with Yellowstone National Park and the surrounding region? The Greater Yellowstone region has a wealth of ice age features that perfectly demonstrate the above explanation:

⇨ The largest caldera in earth's history that probably generated a vast number of earthquakes and subsequent volcanic activity in other parts of the country. The most recent caldera boundary is some 45 miles long and 35 miles wide – a huge cavity out of which poured pyroclastic flows and many cubic miles of lava.

- ⇨ The amount of pyroclastic material from Yellowstone that has been found hundreds of miles from Yellowstone, as far away as Louisiana.
- ⇨ The amount of glacial till found all over the Yellowstone region along with evidence from a Lake Yellowstone that had been much larger in the past.
- ⇨ The amount of glacial erratics found in the northeastern part of the Park.
- ⇨ Evidence of glacial coverage and flooding in the Tetons.
- ⇨ Vast glacial erosion in the Beartooth Mountains. Just a few small glaciers remain in the Beartooth Mountains today. But the evidence that remains from past glaciation is breathtaking.

Following are pictures of this period of glaciation that will be quite obvious. Even though the Bible does not mention an ice age, it is clear that the evidence for it is there. The issue is not *did it occur*, but *when did it occur?* The Bible's timeline has been historically validated through archaeology and passages of Scripture quoted in the New Testament. Further, the Genesis Flood is the only historical mechanism that could adequately explain how an ice age could form along with the formation of the geological features preserved in the Yellowstone region.

A wall of volcanic ash outside of Gardiner, Montana, just outside the north entrance of Yellowstone Park

Glacial horn, called the Bear's Tooth, high in the Beartooth Mountains. Notice the similarity with the famous Matterhorn in Switzerland, also a glacial horn.

Glacial valley in the Beartooth Mountains; an incredibly large classic U-shaped, scoured-out valley which is an obvious sign of glaciation

Glacial kettles and hummocky surface in the Beartooth Mountains – remnants of past glacial occupation

Glacial horns, Pilot and Index Peaks in the Absaroka Mountains, just out the northeast entrance to Yellowstone Park

Glacial erratics in the northeastern part of the Park

Glacial erratics – granite boulders on volcanic terrain! Northeast part of Yellowstone Park

Former Lake Yellowstone in Hayden Valley, Yellowstone Park

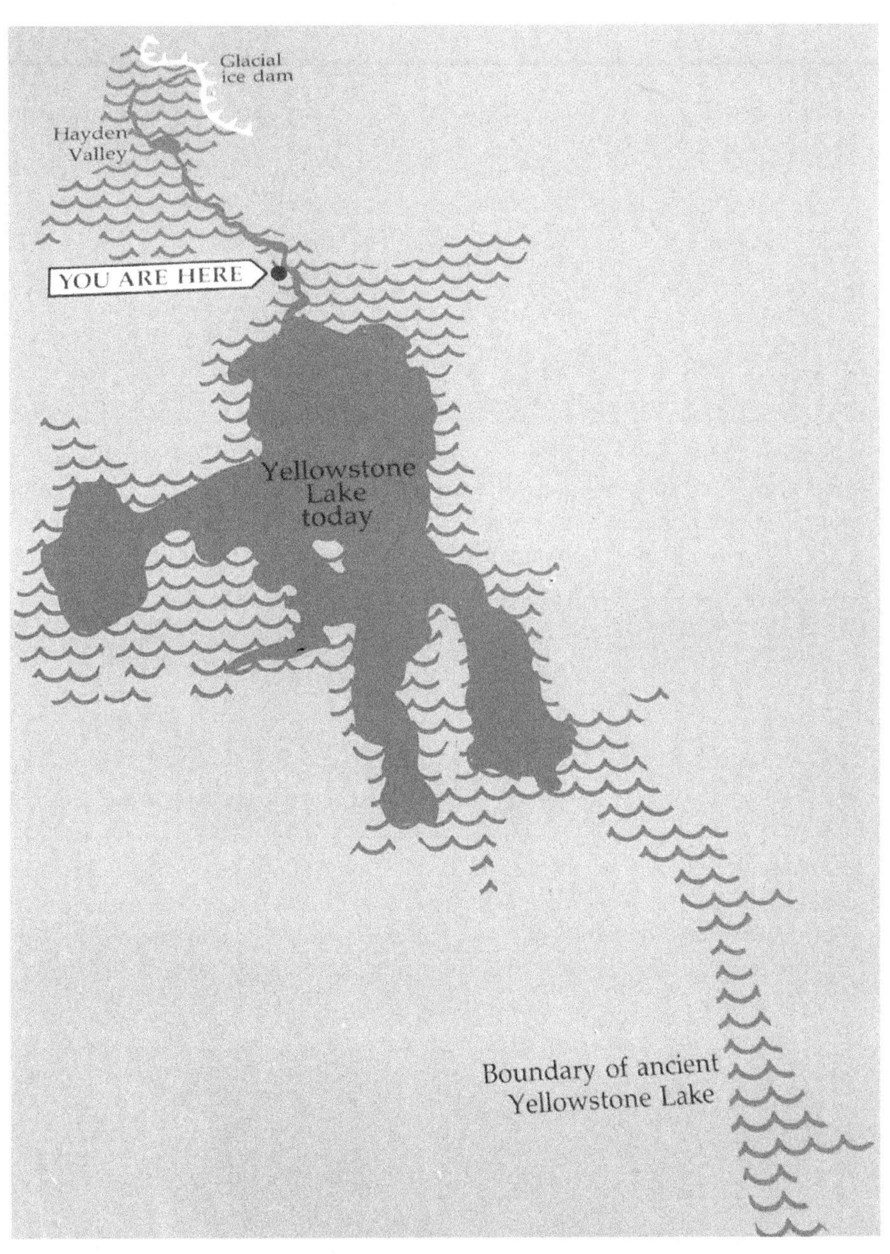

Yellowstone Park sign at the location where the picture above was taken, showing the approximate coverage of former Lake Yellowstone

Extensive water gap at the entrance to Gardiner Canyon, Montana indicating a massive amount of water cutting, either produced by the Genesis Flood or by the catastrophic glacial melt water at the end of the Ice Age. A water gap is a V -shaped canyon cut by a massive amount of water.

The alternating raised grounds in the foreground shows past water levels in the Madison Valley, west of West Yellowstone, most likely from the catastrophic flooding at the end of the Ice Age. The melt water probably came from the melting of the glaciers in the mountains immediately behind me (the photographer).

A glacial cirque that held a glacier that would have produced the melt water.

Tumbled and rounded cobbles in the Madison Valley indicating extensive and violent flooding

Teton Valley where past lake levels have been preserved from glacial flooding and outwash

Glacial valley in the Teton Mountain Range

Glacial cirques in the Teton Mountain Range

The crescent shape above is a glacial moraine in the Lewis and Clark Caverns State Park region, Montana

The Grand Canyon of the Yellowstone is itself a remnant of a massive glacial ice dam burst, similar to the great Missoula Flood ice dam burst.

Although geologists have written extensively about the evidence for an ice age in Yellowstone, its origin remains a mystery. More than 100 ideas have been advanced to explain just how an ice age, the magnitude of the Pleistocene Ice Age, could have begun, let alone ended. Not one of those ideas is completely accepted. A global ice age remains one of the top ten mysteries of modern geology. Only the Genesis Flood with its warmer oceans, volcanic eruptions and huge ash layers around the world could explain such an event in earth history. The Yellowstone Caldera was perhaps the most powerful volcanic eruption in earth history and it would have figured prominently in the onset of a global ice age about 4,500 years ago. Yellowstone National Park is truly an amazing geological place in which to study the geology of the Bible!

Chapter 9
A Brief Road Guide to Yellowstone National Park

This road guide is a brief guide concentrating on the geological features of the Park. It is not meant to be exhaustive, but rather a helpful simple guide so that you can enjoy the geology of Yellowstone Park. The guide will start at The Lewis and Clark Caverns near Bozeman, Montana and proceed throughout the rest of the Park. Mileage and suggested stops are given along with a map of the particular area being looked at. Be sure to check with the Park service, Montana Department of Transportation and the Wyoming Department of Transportation for updated construction and weather reports. Weather can change on a dime! I have seen blizzards in August in The Beartooth Mountains come up without any notice. The Park speed limit is 45 mph. Please stay alert as the designated driver. You cannot drive and sightsee at the same time! I have seen some horrendous traffic accidents with fatalities in Yellowstone because the driver of one of the vehicles involved was not paying attention. Enjoy your trip and trade off driving so that all can get a chance to study this geological wonder!

The Lewis and Clark Caverns

The Lewis and Clark Caverns are located approximately 45 miles west of Bozeman, Montana, and 60 miles northwest from the northwest corner of Yellowstone National Park. The caverns are also notable in that much of the work done to make the cave system accessible to tourists was performed by the New Deal-era Civilian Conservation Corps.

The caverns are part of a vast limestone formation called the Madison Limestone. The Madison and its equivalent strata extend from the Black Hills of western South Dakota to western Montana and eastern Idaho, and from the Canadian border through western Colorado, to the Grand Canyon of Arizona. This is a huge formation of limestone, a sedimentary rock laid down by water and lime-rich mud. It is rich in marine fossils and is over 500 feet thick. The Madison group requires a catastrophic flood explanation for such a tremendous amount of limestone. No shallow sea or localized lake could have formed this much limestone at one time. It is just too big of a limestone formation. The Lewis and Clark Caverns are well worth the short drive into some scenic country. The guided tours take about two hours and descend roughly 600 feet into some spectacular limestone cave formations. Lewis and Clark camped within site of the Caverns in 1802, but did not discover the caves. The cave system was discovered by two non-native Americans in 1882 who immediately recognized the commercial value in giving tours into the caves. Lewis and Clark Caverns State Park is Montana's first state park.

Cave Formation

The typical uniformitarian explanation for cave formation requires thousands of years of slowly penetrating water percolating through limestone and seeping through cracks in the limestone to form a type of limestone in caverns called dripstone (the familiar stalactites and stalagmites). The secular age for most caves is on the order of hundreds of thousands of years based on the present observation of slowly dripping limy water in caves. Of course, if this is true, the Bible cannot be true, as

its chronology is too short. But is there another explanation?

Given the global catastrophic Genesis Flood and its huge amounts of limy water and mud, the formation of limestone caves would be a given. As the limy mud was laid down and then later planed and channelized by the Genesis Flood waters, large caverns would have been carved in a short period of time. As the water continued to drain as mountains were being pushed up, dripstone formation would have been initially much faster. Then as the earth settled down after the Flood, dripstone formation would have been slower, as the amount of water available would have been reduced. Hence, the observation of apparently slow dripstone formation. The secular explanation is simply an application of the belief in uniformitarianism – *the present is the key to the past.*

Just outside the caverns in the parking lot, notice the large limestone wall. Of particular interest is the folding and bending of the rock. How did this happen? Rock does not bend when subjected to pressure. It breaks. Clearly, this deposit was subjected to pressure prior to hardening. This could not have happened over millions of years. This must have happened fairly rapidly after the limy mud was laid down. This phenomenon is common around the world, and is a clear testament to the global Genesis Flood.

Hebgen Lake/Earthquake Lake

If you are planning on being in West Yellowstone or arriving through the West Entrance at West Yellowstone, you should take the short side trip out of West Yellowstone to **Hebgen Lake/Quake Lake** area, sight of the 1959 7.3 massive earthquake – the largest in the Rocky Mountains. The Quake Lake Visitor Center is 24 miles from West Yellowstone.

This area is a testament to the destructive powers of geological forces, and that in a very short amount of time.

> The earthquake measured 7.5 on the Richter scale (revised by USGS to 7.3) and caused an 80-million ton landslide, which formed a landslide dam on the Madison River.... The landslide traveled down the south flank of Sheep Mountain, at an estimated 100 miles per hour (160 km/h), killing 28 people who were camping along the shores of Hebgen Lake and downstream along the Madison River. Upstream the faulting caused by the earthquake forced the waters of Hebgen Lake to shift violently. A seiche, a wave effect created by wind, atmospheric pressure, or seismic activity on water, crested over Hebgen Dam, causing cracks and erosion.
>
> The earthquake created fault scarps up to 20 ft. (6.1 m) high in the area near Hebgen Lake and the lake bottom itself dropped the same distance. 32,000 acres (130 km^2) of the area near Hebgen Lake subsided more than 10 ft. (3.0 m). Several geysers in the northwestern sections in Yellowstone National Park erupted and numerous hot springs became temporarily muddied.[5]

To get there, take US Highway 191 out of West Yellowstone starting at the last stop light at the north end of West Yellowstone. Go north toward Bozeman, Montana on US Highway 191 until you get to the intersection for US Highway 287. Turn left and follow the highway to the visitor center. There is much of geological interest in this area related to the cause and effects of earthquakes. So be sure to take in the various points of interest along the way. If you like, there are clearly marked tourists stops all along the way.

The trip is short and won't take long to see the devastation caused by this earthquake. But, if you want to explore the various tourists' stops, plan to leave a good chunk of the day to do this.

If you have time, take the alternate route back to West Yellowstone by turning right out of the visitor center and traveling on US Highway 287 for a couple of miles to the intersection of US 287 and Montana Highway 87. Turn left and follow the highway past Henry's Lake and around to the south and then east to the intersection of Montana Highway 87 and US Highway 20. Take US Highway 20 straight in to West Yellowstone.

---

5 http://en.wikipedia.org/wiki/Quake_Lake
http://en.wikipedia.org/wiki/Wikipedia:Text_of_Creative_Commons_Attribution ShareAlike_3.0_Unported_License, CC by-SA.

The earthquake caused a large dolomite boulder to give way, which then created the landslide that buried the campground and dammed the valley. This picture shows the attachment point for the boulder.

Quake Lake: This lake was created when the valley was dammed by the landslide. In the distance, you can see the attachment spot for the boulder that gave way.

The Madison River below Quake Lake. The roadway now goes over the dam that was created by the slide.

Collapsed house on Hebgen Lake when a seiche (wave effect) from the earthquake of 1959 hit the shore; remnants are still visible, standing in several feet of water.

You can also see Quake Lake if you are traveling south out of Bozeman, Montana. Simply take US Highway 191 south out of Bozeman until you get to US Highway 287. US Highway 191 takes you through the **Gallatin Canyon**. It is a beautiful drive and you will see lots of geology!

Through the Gallatin Canyon: a very interesting unconformity exists here at this spot just outside of Big Sky. On the left is rock that is identified as basement rock by secular geologists and dated at around 2-3 billion years old. On the right in the picture, just across the highway is the Madison Limestone, around 350 million years old. What happened to all the missing time and history in between?!

An example of gneiss, a metamorphic rock, along US Highway 191

Yellowstone Park: The Grand Loop

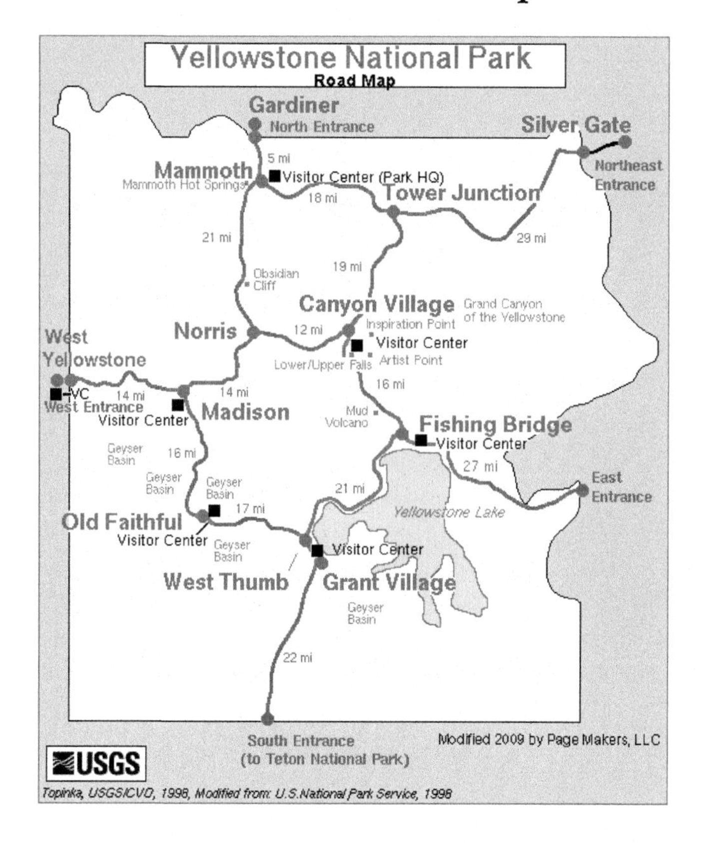

If you study the Park map given to you upon entrance into Yellowstone, you will notice that the Park is crisscrossed by a set of roads that looks like a figure 8. This is called The Grand Loop. If you only have a short time to visit Yellowstone, The Grand Loop is the way to do it. You will get a pretty good picture of what Yellowstone is all about by following this road.

The West Entrance

The west entrance into Yellowstone is just outside of the town of West Yellowstone, Montana. It is 14 miles from West Yellowstone to Madison Junction. The drive into the Park follows the Madison River.

There is a picnic spot about seven miles from the West Entrance. If you want to get a good look at the mountains of volcanic tuff that were produced by the Yellowstone eruption, this is a great spot to see it. Take some time to study this wall of volcanic pyroclastic tuff and try to imagine the heat and wall of ash, volcanic rock and steam moving toward you at incredible speeds. This has been preserved in the wall across the Madison River opposite the picnic area you are standing in.

Tuff is a pyroclastic rock composed of bits and pieces of ash, volcanic rock and rhyolitic glass fused together into a very hard rock. Initially it is erupted within hot steam and lava. As it comes to rest and cools, it is fused together. Now, imagine this amount of very hot ash rolling into this mountain of tuff pictured on the left!

Many feet of tuff are preserved in this wall of ash and volcanic shards of rock that are now welded together into a very hard rock.

The Madison River at Madison is one of my favorite campgrounds with plenty of opportunities to see elk and buffalo.

You should keep your eye open for Trumpeter Swan that frequent the Madison River.

Madison River
Notice the tuff cliff in the background.

The Fires of 1988

Fire is a natural part of the life of the Yellowstone ecosystem. But the summer of 1988 brought fires that many thought would destroy the park.

The summer of 1988 was the driest the park had ever experienced in its history. The mid-late summer storms brought lightning strikes that created fires that simply did not burn themselves out. By the time the fires were extinguished, about 1.2 million acres were scorched; 793,000 (about 36%) of the park's 2,221,800 acres were burned, and 67 structures were destroyed. The debate raged about whether the fire policy for the park was appropriate or adequate.[6]

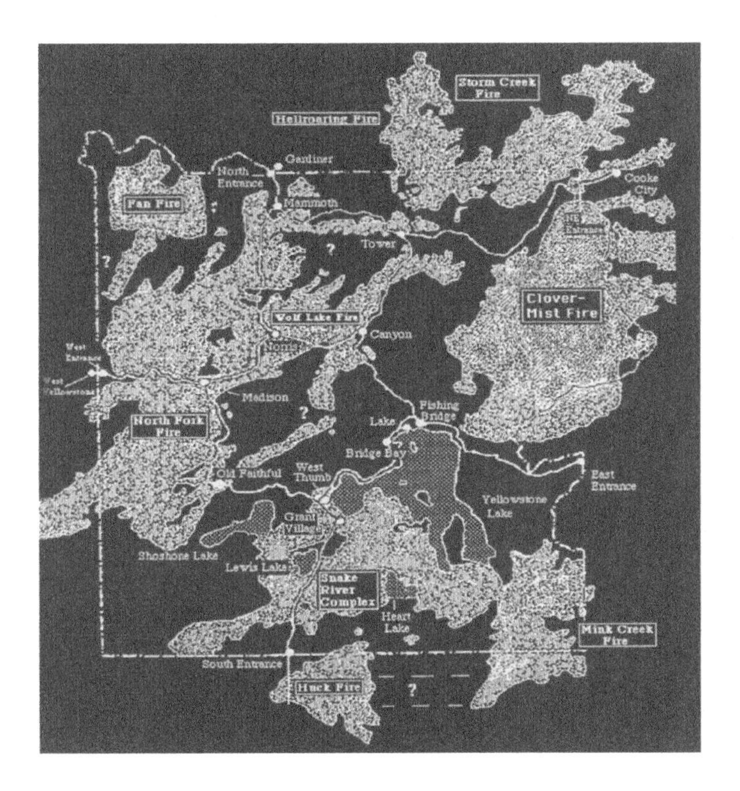

This map shows the extent of the fires late in September, 1988. They finally were extinguished as the snowfalls of September arrived.

[6] Information courtesy of Yellowstone National Park. http://www.yellowstone-bearman.com/yfire.html

Burn area close to Petrified Tree, 1988.

This picture was taken near Obsidian Lake, 1999 (north-northeast of Obsidian Cliff). The young Lodgepole Pines are about 1-2 feet tall. They have created a carpet over the forest floor. Some of the Lodgepole Pines release their seeds from the cones only under intense heat. Consequently, they have in their destruction a resource for the new forest.

The new Lodgepole Pines are beginning to overtake the standing burned trees (snags). This picture was taken near Mammoth Hot Springs, 2010. If you are walking on trails in areas with snags, please be aware that these can fall with no notice.

The area around Madison and north to Norris is a great place to see just how sweeping the fires were. The new growth is not yet at the height of the pre-fire forest, so the burn area is readily recognizable.

North to Norris Geyser Basin

Turn north at Madison Junction to drive toward Norris Geyser Basin. It is a 14-mile drive. As you turn north, you will be following the Gibbon River. About two miles north of the Madison Junction will be **Gibbon Meadows picnic area.** It is a pleasant spot with a couple of interesting features. On the banks of the Gibbon River there is exposed volcanic ash. It is very fine and you may mistake it for dirt.

Also visible on a tree near one of the picnic tables are the claw marks of a grizzly bear. This is all that you want to see of one of these massive bears! When you see this, your curiosity about seeing a grizzly bear will be terminated.

About another three miles north you will come to **Gibbon Falls** on the right. If you like waterfalls, you won't want to miss this one. Take a magnifying glass with you and spend a little time studying the tuff and rhyolite blocks that were taken from the area and used in the construction of the safety walls bordering the Falls. You will get an excellent education in the various minerals that make up the volcanic rock of Yellowstone.

Gibbon Falls

Roughly half way between Madison Junction and Norris Geyser Basin you will come to **Monument Geyser Basin** and **Beryl Spring**. Beryl Spring is one of the hottest springs in Yellowstone, averaging 196 °F! The U.S. Geological Survey Hague party named Beryl Spring in 1883 for the blue-green color, which reminded a party member of the gemstone beryl.

Beryl Spring

The Monument Geyser Basin has no active geysers, but its *monuments* are siliceous sinter (geyserite) deposits similar to the siliceous spires discovered on the floor of Yellowstone Lake. Scientists hypothesize that this basin's structures formed from a hot water system in a glacially dammed lake during the close of the Ice Age. To access Monument Geyser Basin requires a hike up a very steep one-mile trail.

Monument Geyser Basin

After departing the Beryl Spring parking lot, continue north until you come to the turn-off on the right to **Artists Paint Pots**. While taking the walk around the path, notice the encroachment of sinter and water into surrounding trees. This shows the ever-changing nature of the thermal features at Yellowstone. The white coloration on the bottoms of the trees is called *bobby socks.* The Park literature explains these odd-looking features this way:

> *Dead lodge pole pines near some hydrothermal areas look as if they are wearing white anklet socks, at one time called "bobby socks." The dead trees soak up the mineral-laden water. When the water evaporates, the minerals are left behind, turning the lower portion of the trees white.*

Artists Paint Pots and *Bobby Socks*

Norris Geyser Basin

Next stop – Norris Geyser Basin. Norris Geyser Basin is the hottest geyser basin in Yellowstone. One of the reasons for this may be the fact that Norris sits at the intersection of three major faults! It is also one of the most acidic thermal features in the Park. One scientist stated that if Yellowstone were not a National Park, it would be an EPA toxic waste site!

A few notes of **caution** when walking the various trails and boardwalks around Yellowstone – you will no doubt be walking through the steam on several of your walks. Protect your eyewear and glass lenses of your cameras. The steam contains dissolved silica. It can settle on your glass and permanently harden. Once hardened, it is nearly impossible to remove.

Be sure to stay on all Park trails. Do not go off the pathways. You are walking on thin ground in many areas and there have been deaths and severe burns caused by straying off the walkways and falling through the sinter crust.

Also, do not inhale the fumes for a prolonged period. They can make you nauseous. The Park advises that if you feel ill, leave the area immediately.

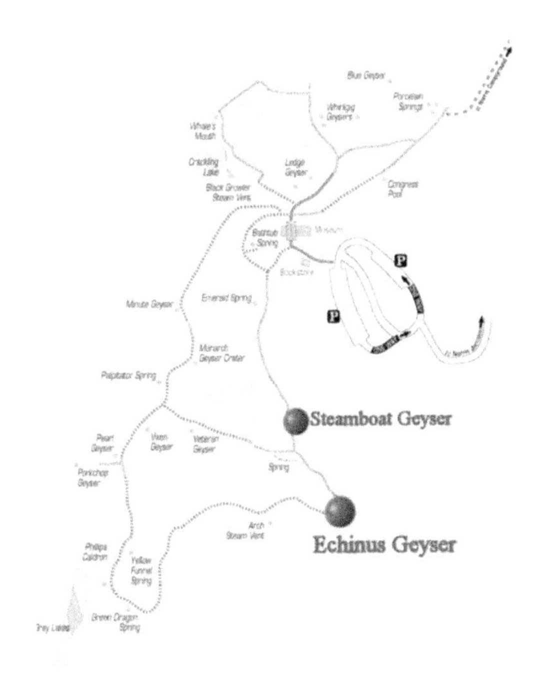

Map of the Norris Geyser Basin

Norris Geyser basin is comprised of two distinct sections separated by a small museum: **The Back Basin**, a 1½ mile walk, and **Porcelain Basin**, a ¾ mile walk. Both are relatively easy, and well worth the effort. Be sure to take a walking-tour map available at each starting point. This is a wonderful way to take in all the sights and smells that remind us of Yellowstone National Park!

Back Basin

Porcelain Basin

Back Basin contains the famous **Echinus and Steamboat Geysers**, rarely seen erupting today.

During the 1980s and 1990s Echinus was quite predictable. I was fortunate enough to see it erupt on every visit. It was spectacular! Today, Echinus is unpredictable. It has a pH of 3, making it extremely acidic.

Echinus Geyser

Likewise, Steamboat Geyser is unpredictable. Most recently, it erupted in 2005, in 2013, and again in 2014. It is the tallest geyser in the world, erupting over 300 feet into the air!

Steamboat geyser erupting in the 1960's, and in 2005

Steamboat Springs as it usually appears.

Depending on what time you visit Yellowstone, you may want to arrive at Norris in the early morning. Around mid-morning the parking lot can become so full that the Park Rangers will close off the parking lot.

Built in 1924, the museum at Norris on a national historic Landmark. It has a nice display on geothermal geology. There is also a bookstore and plenty of restrooms. The Museum of the National Park Ranger is located at Norris Campground, about one mile north of Norris Geyser Basin. It is the original military barracks used during the early days of the Park when the US Army administered the Park.

Twin Lakes

On leaving Norris Geyser Basin continuing north, you will come to Twin lakes. Twin Lakes are a pair of lakes originally given their name in 1879 by Superintendent Philetus W. Norris for their proximity and similar mirrored appearance to one another. North Twin Lake's water comes from an underground spring and runoff from nearby geyser activity. It is very shallow and cannot support a fish population. Despite this, the lake was stocked with whitefish in 1889, and grayling in 1934-1935. In both cases, the lack of suitable spawning habitat led to the demise of the transplanted fish. South Twin Lake's water also comes from springs and geyser activity. And while it is somewhat deeper, it also lacks inlet and outlet channels and therefore lacks suitable spawning habitat to allow for fish reproduction. South Twin Lake was also stocked with whitefish in 1889, as well as 200 rainbow trout in 1933. Between 1934 and 1956, some 450,000 grayling were introduced to the lake. By the mid-1970s, the lake was once again barren. Both lakes have no defined inlet or outlet channels. There have been several studies conducted in an attempt to ascertain where the water from these lakes flows. It is possible that the outlets for these lakes shift from time to time. Both lakes are currently listed as draining into a tributary of Obsidian Creek.

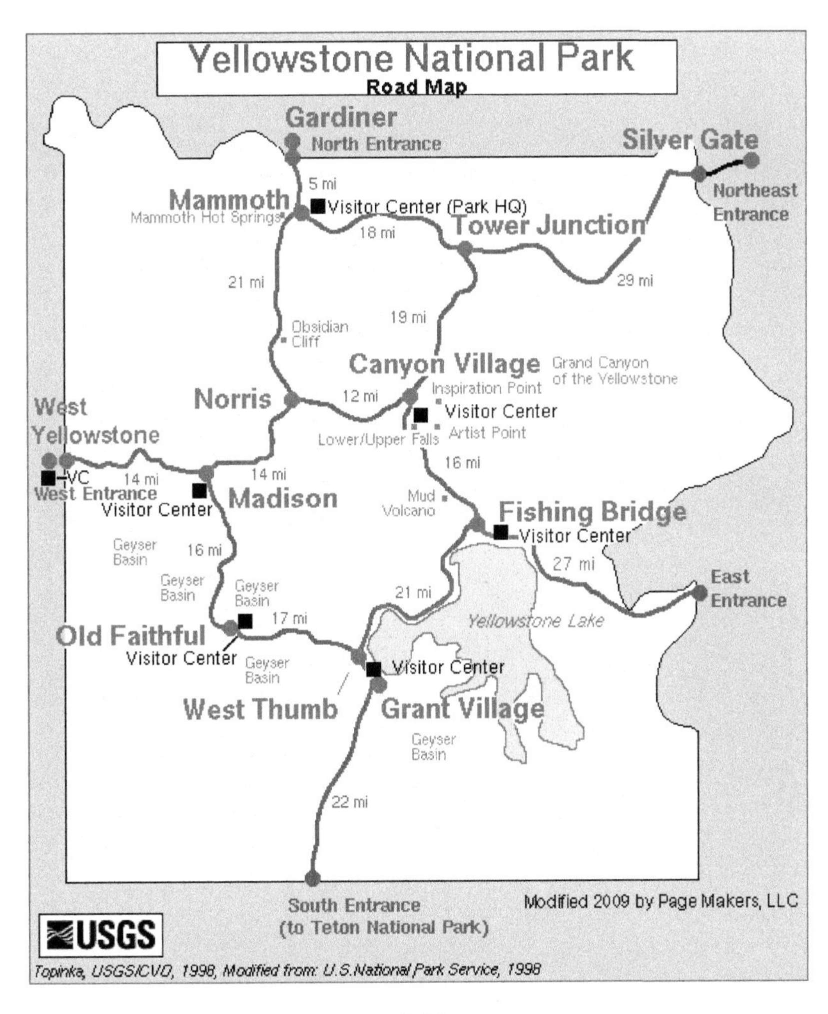

Roaring Mountain

Next stop, just past Twin Lakes and about five miles north of Norris Geyser Basin, you will come to **Roaring Mountain** on the right. Roaring Mountain was named for the numerous fumaroles on its slopes, which during the early 1900s were loud enough to be heard for several miles. The picture above was taken in 1942 by Ansel Adams. Today, only hissing is heard. An earthquake at just the right place and time, however, could once again change this.

Roaring Mountain in 2010

Obsidian Cliff

Continuing north past Roaring Mountain for about 2.5 miles, you will come to **Obsidian Cliff**. Obsidian Cliff was formed from thick rhyolite lava flows. But wait – rhyolite is supposed to be a light-colored rock. Why is obsidian black? It is black because it has a lot of iron that makes it dark colored. Other than the iron, however, obsidian is pure volcanic glass – quartz. Many geologists will say that obsidian has no crystal structure because it formed quickly. But I am not aware of any one who has seen obsidian form. We do not know whether it formed quickly or was simply produced under unusual chemical conditions.

Obsidian Cliff is over five square miles of obsidian flows. (Note: You are not allowed to hike or walk in this area. So be prepared to look as you drive by!) If you will look up the cliff as you slowly drive by, you will notice obsidian columns – the same shape as basalt columns discussed earlier in the book. Lava crystalizes in many shapes and patterns. It is a very interesting rock to collect and study. My collection has more varieties of volcanic rock than any other rock I have collected. It is amazing stuff.

Obsidian Cliff in columnar jointing patterns

Obsidian boulder

In addition to the prohibition to hike or walk in this area, it is also illegal to collect the obsidian – even the stuff that falls onto the road. The Federal Government will fine you heavily for violating this law! It is now officially on the National Register of Historic Places. It was declared a National Historic Landmark in 1996, primarily for its connection to Native American History. The obsidian from Obsidian Cliff, made into arrowheads and other tools, has been found all over the United States, as Native Americans came to look for and trade for this obsidian.

Sheepeater Cliff

A few miles north of Obsidian Cliff, you will come to the turn off to **Sheepeater Cliff**, named for the Sheepeater Indians, a tribe of the Shoshoni Indians who frequented this area and may have lived here. Why are they called *Sheepeater*? You guessed it – they lived on the Rams that were abundant here in those days. The Sheepeater also made bows out of the Rams' horns. They got to be so expert at it that they developed a thriving trade with other Native American tribes.

Sheepeater Cliff is a short drive at the turn-off or a short walk, if you are traveling in a motor home or pulling a trailer. Only cars are allowed to use the road. There are a few picnic tables and a restroom at the site.

When you arrive, you will be treated to a perfect example of basalt columns. Please watch your kids. It is very tempting to climb on the basalt rocks that have fallen through breakage caused by freezing and thawing. But it is dangerous and the columns are weak and could crumble without notice.

Basalt is a dark-colored volcanic rock due to the dark rock-forming minerals, primarily calcium feldspar, pyroxene, biotite, olivine and iron. As the basalt comes to rest during its flow, it will cool and shrink causing symmetrical columns to form into hexagonal and pentagonal patterns.

Basalt lava flow of columnar jointing

Bunsen Peak

As you continue north to Mammoth Hot Springs, the obvious mountain off to the right is **Bunsen Peak**, named after, you guessed it, Robert Bunsen, the inventor of the Bunsen burner. His connection to Yellowstone is not so obvious. The mountain was named after him because of his early work on volcanic geyser theories. Geologists think that as much as 750 feet of ice once covered Bunsen Peak during the Ice Age.

Bunsen Peak looks like a volcano. But it never quite erupted. It is what geologists call a *laccolith.* The word laccolith comes from two Greek words meaning *pond stone.* It is technically magma that has intruded sedimentary layers and then hardened before erupting as a volcano. The rock of a laccolith is typically of high quartz indicating that the magma was viscous and therefore never making it to the surface. Bunsen Peak most likely was forming during the early stage of the Genesis Flood, while being covered by the deposition of sedimentary layers. These sedimentary layers were then removed during the second stage of the Flood that washed them away, exposing resistant magma below. Imagine the amount of sedimentary layers that once covered this area! This begins to give a sense of the power and scope of the Genesis Flood.

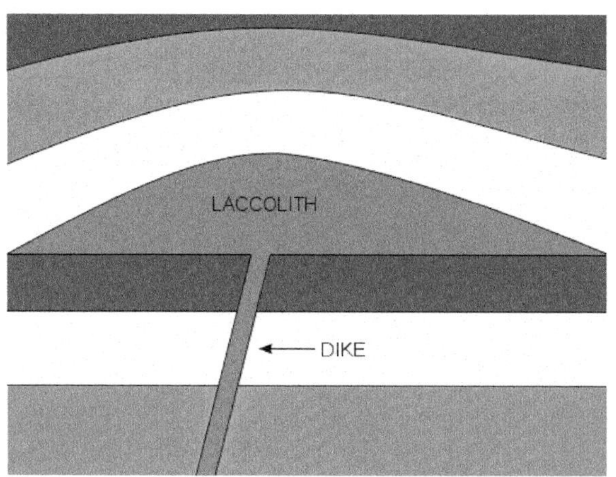

Hypothetical diagram of a laccolith

Continuing past Bunsen Peak, you will arrive at the **Golden Gate Bridge**. On your left is a massive ridge, called Huckleberry Ridge, made out of tuff. It is named, appropriately, the Huckleberry Ridge Tuff. There is plenty of parking here and opposite of the ridge is a small but beautiful waterfall. It is worth the time to stop here and take in the geology of the tuff. Try to get close enough with a magnifying glass to see what is in this pyroclastic material. You will be amazed! Drive slowly across the bridge, noting the height of this ridge above you. Now, imagine this now hardened rock as a massive amount of hot steam and rock rolling through at over a hundred miles an hour. Impressive, isn't it?

During the 1959 Hebgen Lake earthquake, a lot of rock from Huckleberry Ridge was knocked down and closed the bridge.

The Huckelberry Ridge Tuff estimated coverage from the Island Park Yellowstone eruption in Idaho, that produced an estimated 600 cubic miles of pyroclastic material. By comparison, Mt. St. Helens produced just .25 cubic miles of volcanic material.

Mammoth Hot Springs

Your next destination will be **Mammoth Hot Springs**. It now serves as the Yellowstone National Park Headquarters. In the early days of the military administration of the Park, the main military was housed here. It was originally named Ft. Sheridan and then changed to Ft. Yellowstone. Many of the original buildings including officers' quarters are still in use as government buildings.

Mammoth is a unique part of the Park. The thermal activity that occurs here is the same type as in the rest of the park, but it occurs in limestone (a sedimentary rock), instead of rhyolite (a volcanic rock). Hot, acidic waters of carbonic acid rise through limestone below, dissolve it and then carry it to the surface where carbon dioxide is released and then precipitated as a white limestone deposit called travertine. In time microbes and minerals such as iron will give color to the fresh travertine. After the particular spring stops flowing, the travertine will turn to a grayish color.

Mammoth Terraces showing both active and inactive springs. The grayish white travertine signifies that the spring is no longer active in that part of the terrace.

One interesting travertine formation at Mammoth is actually an extinct travertine cone that was once very active. It is called **Liberty Cap** because it reminded the early explorers of the British military caps.

Active hot springs at Mammoth Hot Springs

Toward the North Entrance

As you make your way toward **Gardiner, Montana** from Mammoth, you will descend into Gardiner Canyon through which are scattered tons of rounded cobbles and boulders from extensive flooding probably from the release of rushing melt-water of the glaciers from the Ice Age following the Flood. Notice the rounded cobbles in the streambeds as you travel down the road.

This is an indication of the amount of floodwater that came through here at the close of the ice age. Keep your eyes open for mountain goats making their way across the rocks above.

Mountain Goat above Gardiner, Montana

At the North Entrance: Roosevelt Arch –
For the benefit and enjoyment of the People

Theodore Roosevelt probably did more to advance the National Park idea than any other President. He laid down the cornerstone of the Roosevelt Arch in 1903. The top of the arch is inscribed with a quote from the Organic Act of 1872, the legislation that created Yellowstone as a national park, which reads, *For the Benefit and Enjoyment of the People.*

If you are leaving the Park at this point, you will enjoy a beautiful drive through Paradise Valley on US Highway 89, north of Gardiner. Notice Electric Peak on the left and the glacial out-wash in the form of rounded cobbles and hummocks all along the Yellowstone River. Feel free to pull over at one of the many pullouts and collect all kinds of rock from travertine to volcanic to metamorphic rocks washed out and tumbled during the glacial period ice dam breaches that flowed through this valley.

The Yellowstone River flowing through Paradise Valley

Electric Peak got its name from the numerous lightning strikes the early expeditions experienced climbing this mountain.

Electric Peak, outside of Gardiner, Montana; notice the glacial cirque toward the top and the hummocky surface in the foreground. The hummocky surface consists of loose glacial till that shifted and sank.

Throughout Gardiner Canyon, keep your eyes peeled for glacial moraines and drumlins. Geologists think this area was the furthest north that the glaciers moved during the Ice Age. Drumlins are triangular shaped mounds of glacial till usually covered by foliage that have been left as glacial remnants. As you drive through Paradise Valley heading north out of Gardiner, you will see a number of these on the east side of the highway. On the left side of the highway you will see clear examples of glacial moraines.

Example of a drumlin: the glacier flowed from left to right.

Glacial moraine along the Firehole River, Yellowstone

Glacial Till – Yellowstone Park

East to Tower/Roosevelt

If you are continuing your journey across the north part of the Park, take the exit at Mammoth for **Tower-Roosevelt**, which is a beautiful 29-mile trek. A mile west of Tower-Roosevelt will be a turn-off for the **Petrified Tree**. Only cars are allowed to make the drive all the way to the trail head. If you are in an RV, you will need to park at the small parking lot just after the turnoff and walk to the trail head. It is about a ½ mile walk from the entrance parking area, if you have parked. The petrified tree is but a fraction of the petrified logs buried in the volcanic ash and mudflows of the Absaroka Mountains. The Absaroka Mountains contain an incredible amount of volcanic material. Geologists estimate that this volcanic period erupted around 9,200 cubic miles of volcanic material!

East to the Northeast Entrance and Specimen Ridge

If you continue east toward the Northeast Entrance of Yellowstone Park, you will see signs for **Specimen Ridge**. You can take trails to see excellent samples of petrified trees moved there by the volcanic mudflows of Yellowstone. Thousands of petrified trees are in all kinds of positions as they were transported in and buried in silica-rich ash and

mud. The silica replaced the living cells of the trees and petrified them. How this happened, no one is entirely sure. A special chemical process definitely took place, but what exactly happened is not entirely understood. The hikes into Specimen Ridge are rigorous and require a good portion of your day. It is about five miles, round trip. Do not take this hike alone or without adequate water and first aid supplies!

Specimen Ridge, 1890 – Specimen Ridge 2013 with fossil tree next to a non-petrified tree

In the 1800s because of the uniformitarian framework that had overtaken geology, paleontologists viewed Specimen Ridge as successively buried forests that had grown, died, decayed and were fossilized. They taught that these forests represented 40,000 years of such forests. This destroyed many people's faith in the Bible. The Bible could not possibly be true if these forests were older than what the Bible allowed for the creation of the earth! Then in 1980 it was demonstrated in the eruption of Mt. St. Helens and its subsequent mud flows that trees had been uprooted and moved into all kinds of positions. Some even looked like they had grown in place. After Mt. St. Helens, geologists began to reinterpret Specimen Ridge the same way. Today the Park for the most part has changed its tune! They now say that the trees were likely carried in by volcanic mud flows.

Specimen Ridge is part of the Lamar River Formation, which is part of the Absaroka Volcanic Supergroup. This whole area is a thick accumulation of volcanic rocks that are part of the eruption of stratovolcanoes. Before they were destroyed by erosion, these volcanoes are estimated to have had peaks that rose about 8,000 feet to 10,000 feet above adjacent intermountain valleys.

Tuff boulder in the Lamar Valley, northeast Yellowstone Park; this tuff boulder is a mishmash of pyroclastic material and amygdules. Amygdules seem to have formed as the result of gas bubbles that were filled in with a secondary mineral such as calcite, quartz, chlorite or one of the zeolite minerals.

Depending on location, the Lamar River Formation appears to unconformably overlie older lavas, conglomerates, tuffs, and volcanic breccias of the Sepulcher Formation, Mississippian limestones and dolomites, and Precambrian gneiss. In other words, geologists have a mess!

The Great Unconformity – You can see this feature very clearly as you are getting closer to the northeast entrance, as you travel through the Lamar Valley, going up to Cooke City. The bottom layer which is composed of limestone, is just above the lower tree line. Geologists have dated this as 350 million years old. It is filled with marine fossils. It rests directly on 2-3 billion-year-old basement granite and gneiss (hidden by the trees). The rock that rests directly on the limestone (the top part of the picture) is supposed to be 40 million-year-old Absaroka volcanic rocks - a mixture of tuff, breccia and lava. But the contacts are flat between the basement and limestone, and between the limestone and

Absaroka volcanics, indicating that there must not have been much time that passed for these three events. There is no evidence of erosion events between the contacts. Maybe the Flood explains this mystery in modern geology much better.

The Great Unconformity in Yellowstone Park

Remember that an unconformity represents missing geologic time and material. Essentially what appears to uniformitarian geologists is that 40 million-year-old volcanic rock rests directly on limestone that is supposed to be 350 million years old. This in turn rests directly on Precambrian gneiss and granite, supposed to be 2-3 billion years old. That is about 2 billion years' worth of geologic time that is missing! Where did it go? That is the big question. Some geologists have called this mess, *The Great Unconformity!* The Genesis Flood would account for this formation, having produced the various events rather quickly through catastrophism. In other words, there is no missing time. The whole drive out to the Northeast Entrance is all along this unconformity. In some places, you can see this unconformity very clearly with the 40 million-year-old Absaroka volcanics resting directly on 350 million-year-old Mississippian limestone – and the contact between these two is flat and sharp, indicating there was no erosion of the so-called missing layers.

Along this drive, keep your eyes open for glacial erratics. An erratic is a boulder of different rock than what it sits on. The boulders are made of granite. But the rock they rest on is volcanic. Where did these come from? The only source for the granite is from the Beartooth Mountains, about 50 miles away! They must have been transported in on the glaciers that had covered this area in the past. Further, the boulders are not rounded or tumbled, so they cannot have been deposited by fast moving water. If they had, they would have been smoother and the whole landscape would have been planed.

Glacial erratics cover the ground in the northeastern part of Yellowstone Park.

The rest of the drive out the Northeast Entrance takes you into the Beartooth Mountains for 49 miles of some of the most scenic country in the United States! Be sure to check road conditions before planning your day. Blizzards can spring up suddenly in the Beartooths and the Highway can close quickly.

The Scenic Beartooth Highway

The Beartooth Highway (US Highway 212) runs from Cooke City, Montana to Red Lodge, Montana. It has been called the most scenic drive in America. I have been over that highway many, many times and I enjoy the drive every time.

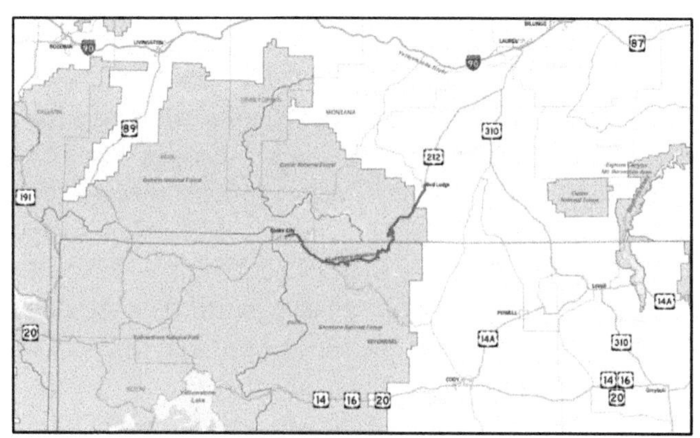

Map of the Beartooth Highway

The name *beartooth* comes from the conspicuous glacial horn at the top of the Beartooth Plateau, called The Bear's Tooth. The Crow Indians called it, *Na Pet Say* – the bear's tooth. It is in a federally protected area called the Absaroka-Beartooth Wilderness, included in the Wilderness Act signed into law by Jimmy Carter in 1978. In 1989 the Beartooth

Highway joined 52 other routes to become a National Scenic Byway. The Beartooth Highway runs over what is referred to in geological circles as the Beartooth Block, a huge mass of uplifted granite and gneiss. Secular geologists label these rocks as some of the oldest rocks on earth. As such they would be basement or foundation rocks. The Biblical view would also call them foundation rocks, probably created during Creation Week, but would place their uplift at some time during the late Genesis Flood.

The name Absaroka is derived from the Native American peoples who called this their home. These people today are referred to as Crow Indians. Their original name was *Absarokee*, a combination of words meaning *a large beaked bird.* In fact, the Crow Indians refer to themselves historically as *the children of the big (large) beaked bird.* This was adopted by early explorers to mean raven. Hence the name ultimately given to them, the Crows. But let's back up a minute. The Crow Indians lived in this land many years before the white explorers and it is interesting to consider what the Crows originally thought about the meaning of their name. The Crows tell the story of a large extinct flying creature in their past history. Could this have been the Pteranodon which was also called the Phoenix by the Sioux? It is an interesting story to ponder.

The Beartooth Mountains display some of the most outstanding glacial features anywhere in the United States and was traditionally considered unconquerable by early explorers. Miners had looked for a way to construct a railroad that would take their ores from Cooke City to Red Lodge, but this never materialized.

Cooke City, Montana 2003; Cooke City remains the gateway to the Northeast Entrance to Yellowstone National Park, although there continues to be some mining nearby. Above the City is Republic Mountain or what is referred to by geologists as The Great Unconformity.

A Brief History of the Beartooth Highway

The Beartooth Highway is part of the Bannock Trail used by Native Americans from 1800 to 1878 as a crossing from homelands in the West to the buffalo herds in the eastern prairies. As the buffalo were hunted to almost extinction in the Idaho country, the Native Americans developed a trail leading to the buffalo pastures of the plains. Hence, the Bannock Trail.

In 1882 General Phil Sheridan became the first Cavalryman to cross the Beartooth Mountains using parts of the old Bannock Indian Trail. He was looking for a shorter route to get from Mammoth Hot Springs, the Park headquarters at the time, to Red Lodge, Montana. Following his successful crossing, people began to propose the building of a highway to enable auto traffic to get into Yellowstone Park much quicker for those travelers coming from the Eastern United States.

Approval for funding the construction of the Beartooth Highway came in 1931 setting aside $2,500,000 for building the highway. The highway was officially dedicated on June 14, 1936 – on time and within budget! It has been called an engineering marvel. Just two lives were lost during its construction. This is amazing considering it was considered impossible to construct a highway over this mountain. The highway today consists of numerous switchbacks and constant rock slides. The highway is open for a brief time between late May and early October. The rest of the time it is buried by as much as 15 feet of snow!

Be sure to re-visit Chapter Eight (The Ice Age in Yellowstone) for some of the spectacular features you will see.

Even in the summer the snowpack from the previous winter can be quite tall!

In addition to being a beautiful drive,
the switchbacks can make it a bit hair-raising!

Glacial valley in the Beartooth Mountains at Beartooth Pass with its classic U shape

The Chief Joseph Highway

About 9 miles east of Cooke City, Montana, US Highway 212 intersects with Wyoming Highway 296. If the Beartooth Highway is closed because of snow or bad weather, there is no need to have all your plans ruined. Take the Chief Joseph Scenic Highway (or Byway). It will generally be open when the Beartooth Highway is not. The highway intersects with Wyoming 120 about 37 miles after getting onto the Chief Joseph Highway. At that point you can either drive southeast into Cody, Wyoming (17 miles) or you can continue northeast driving along the east side of the Beartooth Mountains and then a short jog west into Red Lodge, Montana. It is a wonderful trip.

The Chief Joseph Highway is a breathtaking drive through some of the most spectacular rock formations anywhere on your trip. The route crosses through the Absaroka Mountains, a massive volcanic remnant of Flood/post-Flood volcanism. It gives you a sense of some of the power that must have been unleashed during and after the Genesis Flood!

The highway follows the path that Chief Joseph and the Nez Perce took in their attempts to escape the US Cavalry by going out through Yellowstone Park in 1877.

One of the most fascinating features of this drive is Heart Mountain, visible at the intersection of Wyoming 296 and Wyoming 120. Secular geologists tell us that in the geologic past, a huge piece of the mountain near Cooke City, Montana broke off of the surrounding mountains and literally slid between 25 and 50 miles at speeds of up to 700 miles an hour and came to rest at a point in Wyoming (Heart Mountain). This has been called the Heart Mountain Detachment and it is a humdinger of a geological mystery! Some think it slid by way of lubrication from hot volcanics. At any rate, it was one of the greatest catastrophic land events in earth history! How does a massive piece of a mountain break apart, slide for 25 to 50 miles without breaking up into billions of unrecognizable pieces of rock debris all the way? The only possible way for this to have happened is if the whole affair took place at the end stages of the Flood as the mountains were rising up. The catastrophic event must have taken place under water for lubrication. This is quite a sight and another of the many evidences for the Genesis Flood. *"This is by far the largest rockslide known on land on the surface of the earth and is comparable in scale to some of the largest known submarine landslides."*[7]

The Chief Joseph Highway is filled with scenes of huge masses of uplifted rock and broken rock formations like this one. Pictures just cannot capture the sense of awe and power that were a part of the Genesis catastrophic flood.

Heart Mountain – the 8,100-foot remnant of the famous Heart Mountain Detachment event. It consists of the sedimentary rocks dolomite and limestone broken off of the sedimentary formation, which had rested under the Absaroka volcanic mountains just outside of Cooke City, Montana. It is staggers the imagination to consider that this remnant broke off and slid.

---

[7] Giant rock slab slid on hot lube. New Scientist. 7 May 2005. p. 19. Retrieved 2006-05-17.

South at Tower/Roosevelt

If your journey takes you south at Tower-Roosevelt, you may want to stop at **Tower Fall** and take the short walk to the lookout where you can see the Tower waterfall. If you look across the Yellowstone while at this lookout, you will see a couple of lines of basalt columns. This formation represents another geologic unconformity and a puzzling one. Unbiased perspective would conclude that the volcanic rock at the bottom was laid down first, then the basalt flow, then more volcanic rock followed by another basalt flow and finally the glacial till on top. And, as the contacts among these various geologic rock events are flat, one would conclude that these events occurred rather rapidly and within one larger event. But secular geologists don't see it the same way. Because of their uniformitarian framework, they depend on radiometric dating to help them put the pieces together. But as we have seen earlier in this guide, (1) the dates are unreliable, and (2) the dates ultimately depend on the Geologic Time Scale, which was well in place before radiometric dating came on the scene. So, secular geologists have determined that the volcanic rock on the bottom is about 56 million years old, the first row of basalt columns are around 2 million years old, the gravel on top of the 1st row of basalt is also around 2 million years old, the 2nd row of basalt columns are around 1 million years old and the glacial till on top of it all is around 11,000 years old. Now it should become obvious that there is a whole lot of time and therefore earth history missing! But because they have rejected the Genesis Flood from consideration and therefore a young earth, they account for this missing time simply by stating that the missing time either was never deposited or it eroded away

Tower Fall

The unconformities at Tower Fall: given the historical Genesis Flood, it would make a lot of sense to assume that no time was missing, as each deposit was laid down quickly and successively. You may be wondering what happened to the rest of the formation. Most likely it was washed out by a glacial dam-burst leaving the Yellowstone River as a remnant.

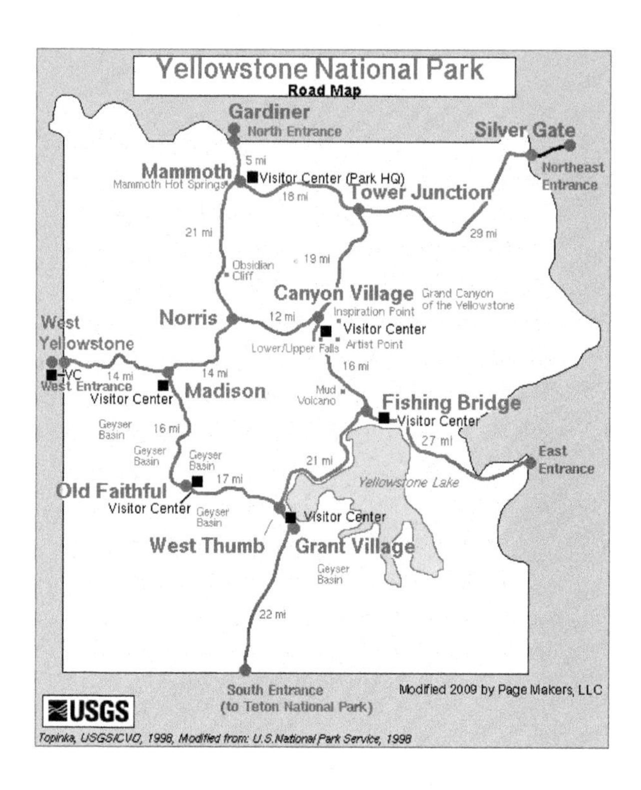

Canyon Village/Grand Canyon of the Yellowstone

The drive heading south out of Tower Junction takes you south 19 miles to **Canyon Village** and the **Grand Canyon of the Yellowstone**. This is one stop you will want to plan the better part of the day to do. There are numerous hikes and trails around Canyon and plenty of shops, restaurants, a new visitor center, a gas station and post office. The main attraction here, of course, is the spectacular Grand Canyon of the Yellowstone. The Canyon is up to 1,200 feet deep and up to 4,000 feet wide. It is a V shaped canyon, which means that it was water-carved.

There is a bit of a geological mystery here. Secular geologists date the canyon as older than the Ice Age. But if that is the case, then, glaciers should have scoured the canyon into a U-shaped canyon. But it's not! And this remains a topic for hot debate among secular geologists. Coming from a Biblical perspective, the Caldera would have erupted shortly after the Flood, laying down the rhyolite lava that would become the future Grand Canyon of the Yellowstone. This was quickly followed by the Ice Age where the entire Yellowstone Plateau was covered in somewhere around 3,000 feet of ice. All this time the volcanic activity would have been going on and maybe this is where the canyon was hydrothermally altered. Geologists tell us that the color of the Canyon is due to the hydrothermal alteration (hot water alteration) because of an abundance of water and volcanic activity. Toward the end of the Ice Age, rapid, catastrophic melting and flooding would have swept over the Yellowstone area, planing it into a plateau and channelizing part of it into the Grand Canyon of the Yellowstone. It's really quite a simple explanation.

The V shaped water-cut canyon of The Grand Canyon of the Yellowstone – this could only have happened during the final stages of the Ice Age. Otherwise any canyon would have been cut into U shaped valleys from glacial movement.

South toward Lake Village/Fishing Bridge

The drive to Lake Village from Canyon Village is 21 miles. As you head south from Canyon, you will be entering **Hayden Valley**. The two places you have the best chance to see lots of animals will be Lamar Valley, out the Northeast Entrance and here in **Hayden Valley**. The buffalo herds have grown significantly through the years while the elk population has shrunk. Many attribute this change to the introduction of the wolf into the Park in 1995. Hayden Valley is the remnant of what used to be Glacial Lake Yellowstone. It was significantly larger in the past due to the melting of the glaciers during the Ice Age. Underlying the grasses is lots of gravel (glacial till) that is not conducive to a flourishing tree population.

Because of this phenomenon, you won't see many trees in this valley.

Hayden Valley, home to many mammals and birds: Glacial Lake Yellowstone once occupied this area.

Sulphur Caldron/Mud Volcano

Following the Yellowstone River, the Grand Loop road travels past the thermal areas that are probably the most dynamic areas of Yellowstone. The first, on the left side of the road is **Sulphur Caldron**. Directly across the highway is **Mud Volcano**. In addition to being very acidic, there have been explosions and upward earth movement in this area. **Sour Creek Dome**, a resurgent dome or volcanic vent, directly behind Sulphur Caldron has been closely watched for several years now. It rises and falls with volcanic activity underneath.

Sulfur is responsible for the peculiar sights, sounds and smells. Iron sulfide paints mud pots and fumaroles shades of brown and gray. Hydrogen sulfide gurgles and hisses and produces a pungent rotten egg smell. Sulfuric acid, twice as acidic as battery acid, cooks the terrain creating a graveyard of skeleton trees.

Sulphur Caldron, an extremely acidic thermal feature

In 1870 when the Washburn Expedition explored the area, the Mud Volcano was much more impressive. It had a tall cone from which it would erupt, shaking the earth, and shooting mud into the air. By 1872, Mud Volcano had ruptured and blown itself apart from all the activity. Today you will not see a cone, but rather a steaming muddy pool framed by a crater-shaped wall that was once the interior of the cone.

Mud Volcano today

Another thermal feature in the area came alive in 1948. **Black Dragon's Caldron**, a mud pot, roared into existence. It blew trees out by the roots and scarred the forested area around it. For several decades, it erupted in 10-20 foot bursts of black mud. Since then it has quieted down and only the remnants remain. It does show, however, that Yellowstone is dynamic and an active remnant of the greater Yellowstone explosions of the past.

Black Dragon's Caldron: quiet now, but it is active, hot and very acidic

We are now at the headwaters of the Yellowstone River. If you are ending your Yellowstone Journey, you will turn east at the intersection just before Lake Village, toward **Fishing Bridge**. Fishing Bridge was named after the original bridge in 1902. It had served as a popular fishing spot up to the 1970s when the Park shut off the bridge to fishing in order to revitalize the Cut Throat trout. The drive will take you out the East Entrance of Yellowstone where you will enjoy beautiful scenery through the Wapiti Valley. Take some of the pullouts and examine some of the volcanic rock. How many different kinds can you find?

Fishing Bridge during the 1950s

Southwest toward West Thumb/Grant Village

As you head south, you will pass **Lake Village** and **Bridge Bay** on your left. It is 21 miles to **West Thumb** from Lake Village. You are now driving parallel to **Yellowstone Lake**, the largest freshwater, high altitude lake in the North America.

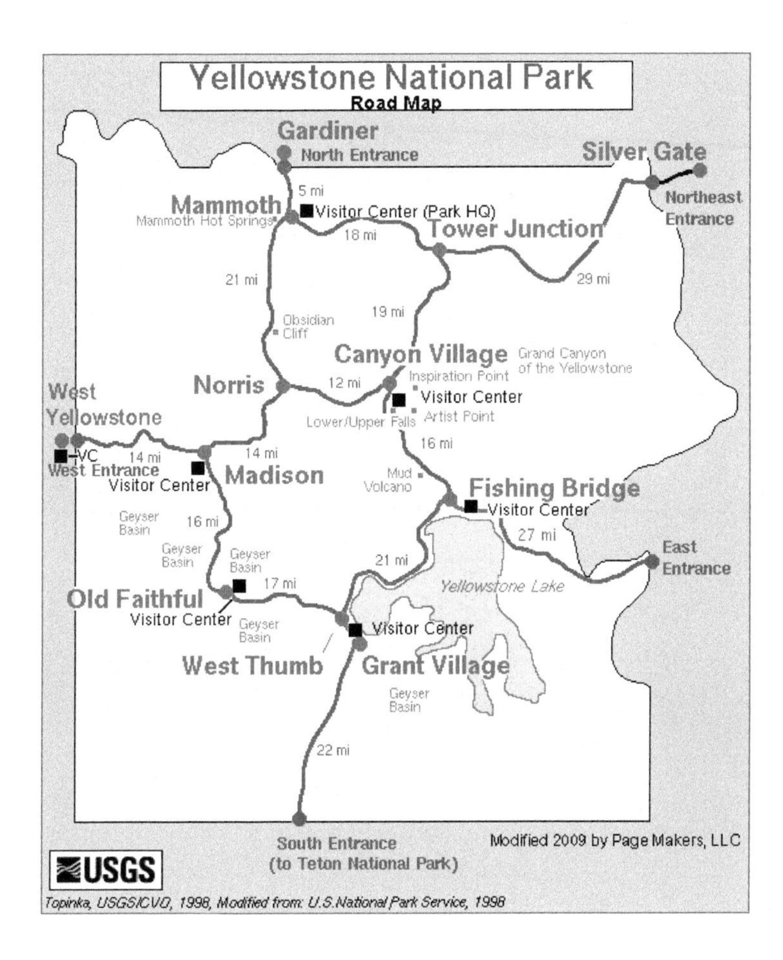

You are at 7,732 feet above sea level. Lake Yellowstone covers 136 square miles with 110 miles of shoreline. During winter ice becomes three feet thick by December and does not thaw out until sometimes as late as early June.

The Lake is considered to be the *final frontier* for exploration. In the last several years, remote–controlled rovers have been used to explore the bottom of the Lake. New thermal features have been discovered including features like *black smokers* found on the ocean bottom. Most of the Lake is the remnant of the Yellowstone Caldera, filled in with the melt-water from the melting of glaciers of the Ice Age.

All along the drive there will be pullouts for more scenic views of the Lake and surrounding mountains. If you have the time, take these drives and take some time to walk along the black sand beaches. The black sand is obsidian tumbled by the waves of Lake Yellowstone.

West Thumb Geyser Basin

You will now be approaching **West Thumb Geyser Basin**. There is plenty of parking and there are pamphlets available for self-guided tours around the Basin. There is also a bookstore that once served as the Ranger's quarters for this area.

West Thumb Geyser Basin is not a large basin, but it has all of the thermal features represented there. It has been more active in the past, but because of the dynamic nature of Yellowstone, could become increasingly more active in the future.

West Thumb Geyser Basin overlooking the West Thumb of Yellowstone Lake

West Thumb is believed to have been a small caldera explosion, the last of the Yellowstone eruptions. It is about the size of Crater Lake in Oregon. The crater has been filled in by the melt-water of melting glaciers of the Ice Age that followed the Genesis Flood.

The heat source for West Thumb is thought to be fairly close to the surface – about 10,000 feet down. Because of the volcanic activity underneath Lake Yellowstone, the shore along West Thumb is shifting slightly downward.

There is another striking example here of how features can change in the park. The thermal features pictured in older photographs where people are seen standing on the rims of the craters are now under water! Among those is **Fishing Cone**. The story is told of people who would stand on the cone rim, catch fish and then cook it by dangling the fish into the cone. Whether that is true or not has not been confirmed. However, there is an interesting old photograph in the Visitor Registration building at Grant Village, a short drive away. It shows a Park Ranger doing exactly what is described above. Below are some photographs documenting the changing nature of Fishing Cone geyser and West Thumb Geyser Basin over the years.

Fishing Cone in 1928

Fishing Cone in 1993, with the lake beginning to envelop it

Fishing Cone in 2010 under water

West Thumb is also a great place to observe the various thermophiles of Yellowstone. Be sure to refer to Chapter Five for explanations of the different microbes.

Some of the beautiful Thermophiles at West Thumb Geyser Basin:
Yellowstone Lake is in the background.

West Thumb Geyser Basin
Spider-like veins in the hot spring are a type of thermophile.

South toward the South Entrance

If you are ending your Yellowstone tour here, you will turn south toward Grand Teton National Park. It is 22 miles to the South entrance to Yellowstone.

West toward Old Faithful

If you are continuing on in your Yellowstone journey, you will turn west toward **Old Faithful**. It is a 17-mile drive to Old Faithful.

The drive to Old Faithful crosses Craig pass, 8,262 feet in elevation. Just past Craig Pass watch for tiny **Isa Lake**. It is the only lake in the United States that drains opposite of what it is supposed to! Normally rivers flowing east of the Continental Divide find their way to the Atlantic Ocean. Rivers flowing west of the Divide end up at the Pacific. The west part of Isa Lake drains to the east, and the east section drains to the west.

Isa Lake with its lily pads: in the spring this lake is filled with the yellow flowers from the lily pads.

Upon leaving Isa Lake, you will have crossed the Continental Divide twice. I believe this may have something to do with the phenomenon concerning Isa Lake.

After crossing the bridge at Isa Lake, you will continue to proceed towards Old Faithful. About three miles east of Old Faithful there will be a parking lot for the trailhead to **Lone Star Geyser**. As is true with many of Yellowstone's geysers and waterfalls, they must be accessed by hiking. Lone Star Geyser is a cone type geyser with eruptions reaching as high as 40 feet. It was in 1882 that the Geyser received its current name. While working in the Upper Geyser Basin, workers for the Northern Pacific Railroad came across this geyser and thought that they had been the first to discover it. But the Hayden Geological Survey of 1872 had made notes of it and named it Solitary Geyser. The 1882 name, however, has stuck. Lone Star Geyser erupts about every three hours, lasting around 30 minutes. The Geyser's cone is made from siliceous sinter (geyserite) formed from repeated eruptions of silica super-saturated hot spring water. The trail to Lone Star Geyser is an old Forest Service road open to hiking and bicycling only. The hike is a five-mile round trip trek. Watch for the brown sign marking the trailhead.

Upper Geyser Basin

It is about two miles from the parking lot for Lone Star Geyser to **Old Faithful** – your next major stop. You will be entering the **Upper Geyser**

Basin. I know it may seem strange that this basin is called the UPPER Geyser Basin, because you are south of the Midway and Lower Geyser Basins. But it makes sense if you understand that you are moving away from Yellowstone Lake, which is the main source of water for the various rivers in Yellowstone Park. Upper Geyser Basin is the closest of the three geyser basins to Yellowstone Lake.

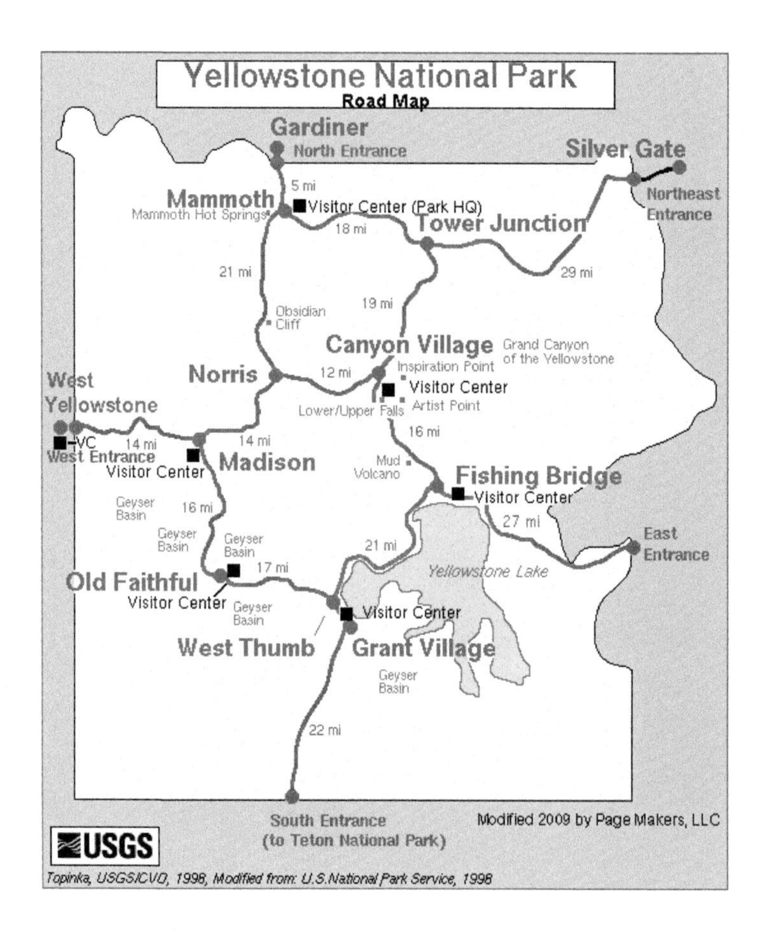

Old Faithful

You could take a few days to walk the trails and see the various thermal features of not only Geyser Hill (home to Old Faithful), but also of Biscuit and Black Sand Basin. Old Faithful is not the only thermal feature in this area. The Upper Geyser Basin has the highest concentration of geothermal features in the Park. There are five major geysers in the Upper Geyser Basin, including, **Old Faithful**, **Castle Geyser**, **Grand Geyser**, **Daisy Geyser**, and **Riverside Geyser**.

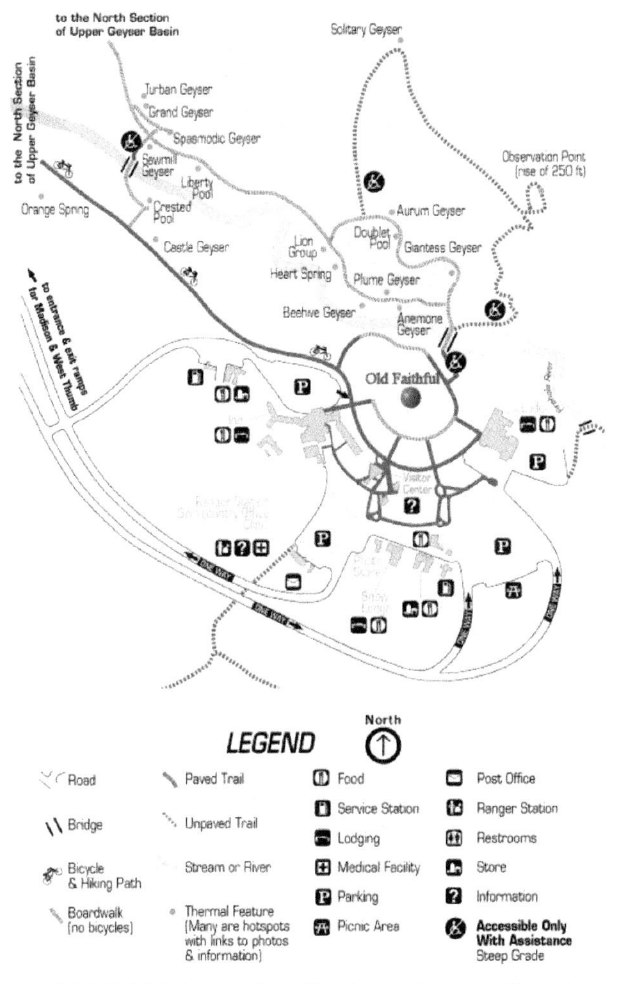

Old Faithful map; it is like being in a small city!

Old Faithful – the most popular geyser

The Upper Geyser Basin also includes two lesser basins called, **Biscuit Basin** and **Black Sand Basin**. Biscuit-like sinter deposits once lined the edge of **Sapphire's** crater, and in the 1880s it received its name for the knobby formations that reminded folks of biscuits. The 1959 Hebgen Lake earthquake caused **Sapphire** to erupt, breaking and dislodging the formations. The name *Black Sand* comes from the crushed Obsidian (volcanic glass) that is present around the geyser basin. The frequent eruptions of Cliff Geyser and the colorful waters of the Sunset Lake thermal feature are a couple of the natural wonders you will see at Black Sand Basin.

Sapphire Pool in Biscuit Basin

The tree-covered hills you see surrounding the **Upper Geyser Basin** are all volcanic, consisting of different rhyolite lava flows, basalt flows and pyroclastic tuff. Geologists estimate that there have been as many as 100 major and minor volcanic eruptions in the Yellowstone area! That is why there appears to be a mish-mash of volcanic rock and debris covering the Park. On top of that, the Ice Age further changed the landform features and moved rock within, into and out of the Park! Sorting it all out continues to be a monumental task.

From a Biblical perspective, the main event which sparked all of these volcanic events, was the Flood. God's intent, mentioned in Genesis chapters 6-8, was to totally destroy man and every living land animal from off the face of the earth. The Flood was catastrophically destructive to say the least! Not since that mind-boggling event have there been such huge volcanic explosions and lava flows. The volcanic explosions we see today are destructive, but not to the degree that Yellowstone, Crater Lake in Oregon, La Garita Caldera in Colorado and the Mammoth Lakes region in California have been.

Sawmill and Grand Geysers erupting in Upper Geyser Basin, 2014

Grand Geyser erupting in 2009

Old Faithful – the icon of Yellowstone; 1877 and the present

Daisy Geyser in the Upper Geyser Basin

The fantastic geyser, Riverside Geyser; it is one of the geysers that you should not miss on your walk around the Upper Geyser Basin. Just check expected times of eruption at the Visitor Center.

Besides geysers, the Upper Geyser Basin hosts several hot springs. Among the favorites is **Morning Glory Pool**. The wife of Assistant Park Superintendent, Charles McGowan, named the pool in 1883. She called it *Convolutus,* the Latin name for the morning glory flower, which the spring resembles. By 1889, the name Morning Glory Pool had become common usage in the park. The distinct color of the pool is due to bacteria, which inhabit the water. On a few rare occasions the Morning Glory Pool has erupted as a geyser, usually following an earthquake or other nearby seismic activity.

Several of Morning Glory's plumbing system conduits have been clogged due to objects being thrown in by tourists over the years, reducing the hot water supply, and in turn altering the overall appearance of the pool. I can remember as a kid seeing all the coins that had been tossed into the Pool by tourists throughout the years. Several attempts by park officials to artificially induce eruptions to clear the pool of debris and clear blocked entryways have been met with mixed results. An interpretive sign, placed near the pool by the park service, discusses the damage caused by ignorance and vandalism and suggests that Morning Glory is becoming a Faded Glory. Some of the vibrant colors are caused by light reflection on the geyserite in addition to the thermophiles that inhabit the pool. So, on sunny and cloudy days, the pool will seem to change in its color patterns.

Morning Glory Pool

One of the short hikes you can take from Biscuit Basin is to **Mystic Falls**, a 70-foot cascading-type water fall along the Firehole River. It is well worth the hike to see this out of the way water fall.

Mystic Falls

North toward Midway Geyser Basin and Grand Prismatic Spring

About four miles north of the Upper Geyser Basin you will come to **Midway Geyser Basin**, located along the Firehole River. The main attractions here are **Grand Prismatic Spring** and the colorful hot spring run-offs into the Firehole River. The colors generated by the thermophile bacteria are fantastic! Grand Prismatic Spring has long been the laboratory for studying the habits of the many kinds of bacteria here. If you want some interesting reading on this subject, get the Yellowstone Park's Seen and Unseen. It is a great paper-back book that discusses in detail the various types of thermophiles and acidothermophiles that inhabit the hot waters of Yellowstone Park.

Run-off into the Firehole River – Midway Geyser Basin; colors are due to the thermophile bacteria that inhabit and thrive in the hot springs of Yellowstone Park

North toward Lower Geyser Basin

After leaving Midway Geyser Basin, just a short drive north of Midway, you will see the entrance to the **Lower Geyser Basin**. There is a short drive you should take called **Firehole Lake Drive**. Many of the thermal features are easily observable right from your car along this drive. **Great Fountain Geyser** is located along this drive. The geyser erupts every nine to 15 hours. Great Fountain's maximum height ranges from about 75 feet to over 220 feet. The duration of the eruption is usually about one hour but durations of over two hours have been seen. The kind and length of an eruption affects the interval that will elapse before the next eruption begins.

Great Fountain Geyser

A very interesting thing happened along Firehole Lake Drive in 2014. The ground underneath got so hot that the asphalt began to melt in places. The road was closed for a short period of time. But it did cause some increased interest inside and outside of Yellowstone Park!

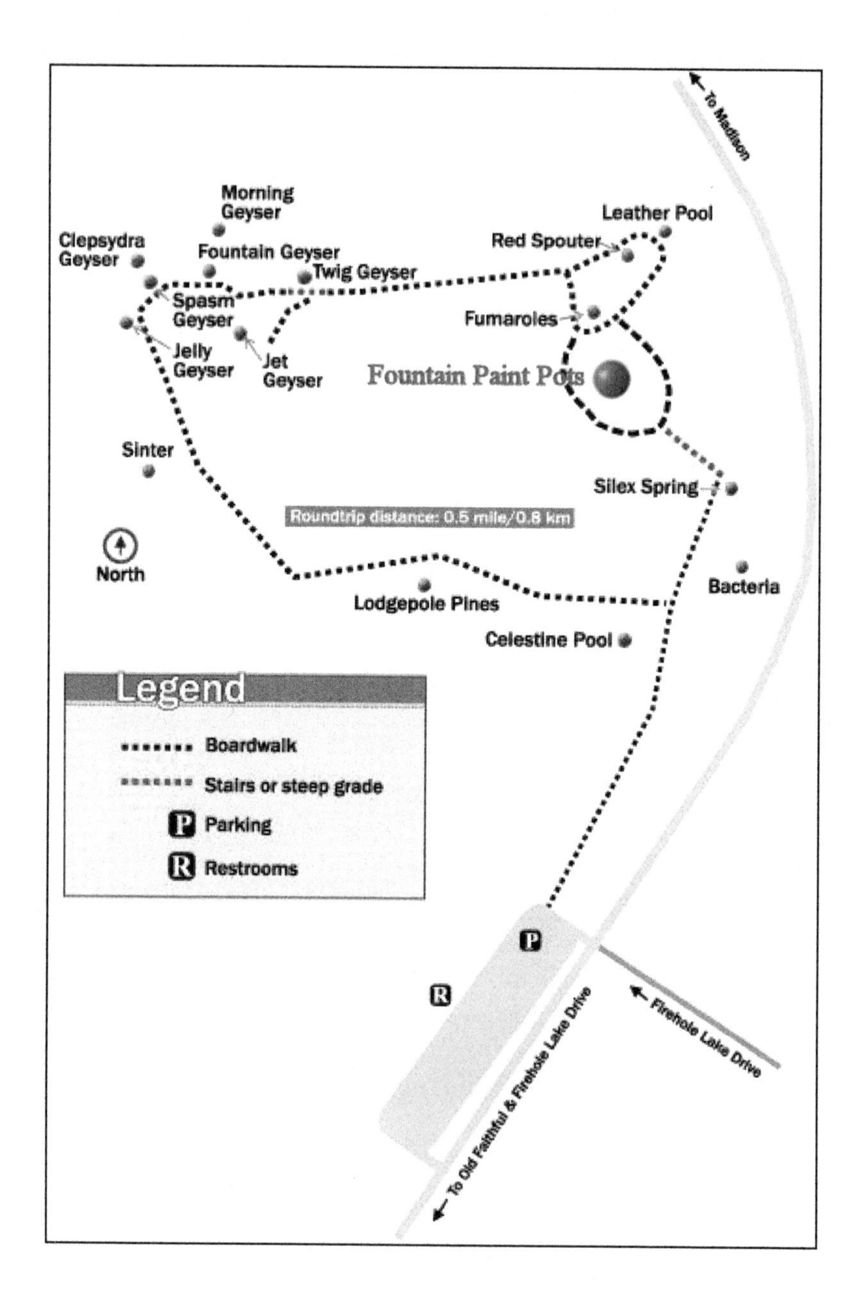

Fountain Paint Pots

Also located within the Lower Geyser Basin is the **Fountain Paint Pots** area. The fascinating Fountain Paint Pot area in Yellowstone National Park contains all four types of thermal features that are popular to view - geysers, hot springs, mud pots and fumaroles. Walk the half-mile boardwalk nature trail to see it all. Visitor pamphlets are available for free if you return them at the end of your tour. The Fountain Paint Pots is named for the reds, yellows and browns of the mud in this area. The differing colors are derived from oxidation of the iron in the mud as well as the thermophile bacteria that inhabit the 150 degree plus temperatures of the water. As with all hot springs, the heat in the caldera forces pressurized water up through the ground, which is expelled here. Also, rising gasses, including carbon dioxide, cause the bubbling action. This picture is of one of the interesting mud pots in the Lower Geyser Basin; the smell is due to the interplay between thermoacidophiles, sulfur, sulfuric acid and sulfur dioxide gas. The acid breaks down the geyserite, converting it to clay. When mixed with the hot water, the clay turns to mud. Rising carbon dioxide gas causes the bubbles to appear like the mud is boiling.

Red Spouter, Fountain Paint Pots

Silex Spring in Lower Geyser Basin, part of the Fountain Paint Pot area

Overflow of hot water from Silex Spring located in the Lower Geyser Basin. The colors are due to a combination of the bacteria and iron oxidation

Around two miles north of the Lower Geyser Basin, you will come to **Fountain Flat Drive**. Watch for the exit to the left and across a small bridge. The picnic area here is a nice serene spot next to the Firehole River. You can sit on the bank of the river and dangle your feet in the cool waters of the Firehole.

Another great waterfall to hike to starting from Fountain Flat Drive is **Fairy Falls**. It is an easy and pleasant hike back to this little-visited waterfall. Fairy Falls is a 197-foot waterfall along Fairy Creek.

Fairy Falls – accessed by way of either Fountain Flat Drive or just beyond Grand Prismatic Spring in Midway Geyser Basin.

North to Madison Junction

From here it is about four miles back to Madison Junction and the end of your tour. Yellowstone Park – what an amazing place and a great place to study Biblical geology. I hope you have enjoyed your tour!

Patrick

Picture Credits

Title page: Photo courtesy of Brad Dennis.

Chapter One

Yellowstone National Park sign: Photo courtesy of Brad Dennis, 6. John Colter: Painting by Alfred Jacob Miller. Found at http://www.yellowstonegate.com/2012/09/early-yellowstone-explorer-john-colter-runs-for-his-life-1807/, 6. Colter's Hell Trail: http://piccola77.blogspot.com/2012/09/wy.html, 7. Article: Courtesy of NPS. Found at http://www.nps.gov/features/yell/slidefile/history/articlesillustrations/Images/03137.jpg, 7. The Hayden Expedition: Public Domain. Found at http://en.wikipedia.org/wiki/Hayden_Geological_Survey_of_1871#mediaviewer/File:Hayden_PackTrain_1871_jwh00114.jpg, 8. William Henry Jackson: Public Domain. Found at http://www.nps.gov/features/yell/slidefile/history/jacksonphotos/Images/02932.jpg 8. Thomas Moran: Photo by Napoleon Sarony. Public Domain. Found at http://en.wikipedia.org/wiki/Thomas_Moran#mediaviewer/File:Portrait_of_Thomas_Moran_by_Napoleon_Sarony.jpg, 8. Thomas Moran's Grand Canyon of the Yellowstone, 1871: Public Domain. Found at http://fr.wikipedia.org/wiki/Thomas_Moran#mediaviewer/File:Thomas_Moran_-_Grand_Canyon_of_the_Yellowstone.jpg, 9. Jackson's photograph of Tower Falls, 1871: Public Domain. Found at http://www.yellowstonegate.com/2013/05/naming-tower-fall-np-langford-1870-2/, 9. Yellowstone Map: Public Domain, found at http://en.wikipedia.org/wiki/Yellowstone_National_Park#mediaviewer/File:Yellowstone_1871b.jpg, 10. Poached bison heads: Photographer unknown. Photo courtesy NPS. Found at http://www.nps.gov/features/yell/slidefile/history/1872_1918/military/Images/16058.jpg, 10. Post Exchange at Ft. Yellowstone: Photographer unknown, courtesy NPS, found at http://www.nps.gov/features/yell/slidefile/history/1872_1918/military/Images/02990.jpg, 11. Military outpost: http://www.flickr.com/photos/yellowstonenps/12679244033/, https://creativecommons.org/licenses/by/2.0/legalcode, CC by 2.0, 11. Union Pacific Railroad Poster: Public Domain. Found at http://en.wikipedia.org/wiki/File:Union_Pacific_Yellowstone_National_Park_Brochure_(1921).JPG, 12. Union Pacific Railroad: http://nationdivided.wikispaces.com/First+Transcontinental+Railroad, http://creativecommons.org/licenses/by-sa/3.0/legalcode, CC by SA 3.0, 12. Advertisement: Pubic Domain. Found at http://en.wikipedia.org/wiki/File:Stagecoach_at_Mammoth_Hotel.jpg, 12. Train Station: Public Domain. Found at http://historyinphotos.blogspot.co.uk/2014/01/old-west.html, 13.
Early car: Public Domain. Found at http://www.yellowstonegate.com/wp-content/uploads/2012/05/merrycarynp3b.jpg, 13. Poster: Public Domain. Found at http://en.wikipedia.org/wiki/Yellowstone_National_Park#mediaviewer/File:Yellowstone_Natl_Park_poster_1938.jpg, 15.

Chapter Two

Carbon atoms: Image by Vicki Nurre, 17. Periodic Table: http://the-operator-theory.wikispaces.com/A+periodic+table+of+periodic+tables, http://creativecommons.org/licenses/by-sa/3.0/legalcode, CC by-SA 3.0, 18. Radioactive atom: http://chemistry3group.wikispaces.com/Uses+of+Radioactive+Decay, http://creativecommons.org/licenses/by-sa/3.0/legalcode, CC by-SA 3.0, 19. Carbon-14: Image by Vicki Nurre, 19. Radioactive decay: Image by Vicki Nurre, 20. Decay chains: Image by Vicki Nurre, 21. Geologic Time Table: http://www.bing.com/images/search?pq=geologic+time+&sc=8-14&sp=2&sk=IA1&q=geologic+time+scale+chart&qft=+filterui:license L2_L3&FORM=R5IR42#view=detail&id=0548D46FA426B2733D713B5A16EE55BFDB889547&selectedIndex=0, 22.Mt. St. Helens Dome: Public Domain. USGS. Found at http://mythphile.hubpages.com/hub/volcano-glossary, 24. Mt. Ngaruhoe: http://www.bing.com/images/search?pq=mt.+ngauruhoe&sc=7-12&sp=-1&sk=&q=mt.+ngauruhoe&qft=+filterui:license-L2_L3&FORM=R5IR42#view=detail&id=B14F9FB532C0724F11E09FDE6FA27081EAF84E1D&selectedIndex=2, https://creativecommons.org/licenses/by-sa/2.0/legalcode, CC by-SA 2.0, 25. Lava flow: Photo by Vicki Nurre, 25. Mini photo: Photo by Vicki Nurre, 25.

Chapter Three

James Hutton: Public Domain. Found at http://es.wikipedia.org/wiki/James_Hutton, CCA-S 3.0, 28.
Charles Lyell: Public Domain. Found at http://commons.wikimedia.org/wiki/File:Charles_Lyell_by_David_Octavius_Hill,_c1843-47.jpg, 28. Mini photos: Photo by Vicki Nurre, 30. Mini photo: Photo by Vicki Nurre, 31. Mini photo: Photo by Vicki Nurre, 33. Mini photo: Courtesy of Brad Dennis, 34. Mini photo: Photo by Vicki Nurre, 37. Mini photo: Photo by Vicki Nurre, 40. Mini photo: Photo by Vicki Nurre, 43. Mini photo: Photo by Vicki Nurre, 45. Mini photo: Photo by Vicki Nurre, 46. Mini photo: Photo courtesy of Brad Dennis, 47. Yellowstone Caldera: Photo by Patrick Nurre, 48. Beartooth Mountains: Photo by Patrick Nurre, 48. Mini photo: Photo by Vicki Nurre, 48.Lone Peak: Photo by Patrick Nurre, 49. Summit of Lone Peak: Photo by Patrick Nurre, 49. Limestone Deposits: Photo by Patrick Nurre, 50. Folded Sedimentary Rock: Photo by Vicki Nurre, 50. Beartooth Butte: Photo by Patrick Nurre, 51. Limestone deposits: Photo by Patrick Nurre, 52. Madison Limestone: Photo by Patrick Nurre, 52. Limestone Mountain: Photo by Patrick Nurre, 52. Limestone wall cut-out: Photo by Vicki Nurre, 53. Author with vug: Photo by Vicki S. Nurre, 53. Dogtooth Crystals: Photo by Vicki S. Nurre, 53. Fossil Ripple Marks: Photo by Patrick Nurre, 54. Mt. Everts: Photo by Patrick Nurre, 54. Huckleberry Ridge Tuff: Photo by Patrick Nurre, 55. Petrified Tree: Photo by Vicki Nurre, 56. Mini photo: Photo courtesy of John Meyer, 56. Principles of Geology: Found at http://abhsscience.wikispaces.com/Charles+Lyell, http://creativecommons.org/licenses/by-sa/3.0/legalcode, CC by-SA 3.0, 57. Mini photo: Photo courtesy of Brad Dennis, 57.

Chapter Four

Map: http://en.wikipedia.org/wiki/Yellowstone_Caldera#mediaviewer/File:HotspotsSRP_update2013.JPG, http://creativecommons.org/licenses/by-sa/3.0/legalcode, CC by-SA 3.0,58. Yellowstone Map:

volcanoreport.wikispaces.com, http://creativecommons.org/licenses/by-sa/3.0/legalcode, CC by-SA 3.0, 59. Crater Rim: Photo by Patrick Nurre, 60. Caldera Explosion: https://geol105naturalhazards.voices.wooster.edu/page/2/, http://creativecommons.org/licenses/by-sa/3.0/legalcode, CC by-SA 3.0, 61. Rhyolite Lava: Photo by Patrick Nurre, 61. Basalt Lava: Photo by Patrick Nurre, 62. Mt. Shasta and Shastina: Photo by Vicki Nurre, 62. Earthquake Map: http://pubs.usgs.gov/fs/2005/3024/images/fs2005-3024_fig_03_large.jpg', 63. Sour Creek Resurgent Dome: Photo by Patrick Nurre, 64. Mini photo: Photo by Vicki Nurre, 64. West Thumb Geyser Basin: Photo by Patrick Nurre, 65.

Chapter Five
Riverside Geyser: Photo by Vicki Nurre, 66. Crystal Geyser: Public Domain. Found at http://en.wikipedia.org/wiki/Crystal_Geyser#mediaviewer/File:Eruption_of_Crystal_Geyser.jpg, 67. Beehive Geyser: Photo by Vicki Nurre, 68. Geyser Diagram: http://www.nps.gov/yell/naturescience/cone_geyser.htm, 69. Siliceous Sinter: Photo by Vicki Nurre, 70. Castle Geyser: Photo by Vicki Nurre, 70. Hot Spring: Photo by Vicki S. Nurre, 72. Grand Prismatic Spring: Photo by Vicki Nurre, 72. Fumeroles: Photo by Patrick Nurre, 73. Roaring Mountain: Photo by Vicki Nurre, 73. Mud Pots: Photo by Patrick Nurre, 74. Mud Volcano: Photo by Patrick Nurre, 74.

Chapter Six
Prokaryote: Diagram by Mariana Ruiz Villarreal. Public Domain. http://en.wikipedia.org/wiki/Bacterial_cell_structure#mediaviewer/File:Average_prokaryote_cell-_en.svg, 74. Tree of Life: http://en.wikipedia.org/wiki/Phylogenetic_tree, http://creativecommons.org/licenses/by-sa/3.0/legalcode, CC by-SA 3.0, 75. Microbes: https://worldofbiology.wikispaces.com/file/view/diff_bacterial_cells.jpg/34813715/diff_bacterial_cells.jpg, http://creativecommons.org/licenses/by-sa/3.0/legalcode, CC by-SA 3.0, 75. Carolus Linnaeus: Public Domain. Found at http://upload.wikimedia.org/wikipedia/commons/thumb/c/c0/Carolus_Linnaeus.jpg/200px-Carolus_Linnaeus.jpg, 77. Grand Prismatic Spring: Photo by Vicki Nurre, 78. Grand Prismatic Spring: Photo by Vicki Nurre, 79. Grand Prismatic Spring: Photo by Vicki Nurre, 79. Chromatic Pool: Photo by Vicki Nurre, 80. Upper Geyser Basin: Photo by Vicki Nurre, 80. West Thumb Geyser Basin: Photo by Vicki Nurre, 80. Upper Geyser Basin: Photo by Vicki Nurre, 81. Microbes in West Thumb Geyser Basin: Photo by Vicki Nurre, 82. Mammoth Hot Springs: Photo by Vicki Nurre, 82. West Thumb Geyser Basin: Photo by Vicki Nurre, 83. Firehole River: Photo by Vicki Nurre, 83. pH Scale: Image by Vicki Nurre, 84. Mud Volcano, Dragon's Caldron, Sour Creek Dome: Photos by Patrick Nurre, 85. Norris Geyser Basin: Photo by Patrick Nurre, 85. Mini photo: Photo by Vicki Nurre, 85.

Chapter Seven
Limestone, Big Sky: Photo by Vicki Nurre, 86. Electric Peak: Public Domain. Found at http://en.wikipedia.org/wiki/Gallatin_Range#mediaviewer/File:Electric-peak-trees.jpg, 87. Petrified Logs: Photo by Greg Willis. Found at http://www.flickr.com/photos/gregw66/3946943581/, https://creativecommons.org/licenses/by-sa/2.0/legalcode, CC by-SA 2.0, 87. Beartooth Mountains: Photo by Vicki Nurre, 88. Absaroka Mountain Range: Photo by Patrick Nurre, 88. Madison Range: Photo by Patrick Nurre, 89. White Cliffs of Dover: Photo by Remi Jouan, found at http://en.wikipedia.org/wiki/White_Cliffs_of_Dover#mediaviewer/File:Douvres_(6).JPG, http://creativecommons.org/licenses/by-sa/3.0/legalcode, CC by-SA 3.0, 90. Light-colored minerals: Photo by Patrick Nurre, 91. Dark-colored minerals: Photo by Patrick Nurre, 91. Minerals: Photos by Heidi Ann Noggle, used by permission, 92. Granite: Photo by Patrick Nurre, 92. Minerals: Photos by Heidi Ann Noggle, used by permission, 92. Gneiss: Photo by Patrick Nurre, 93. Gneiss: Photo by Patrick Nurre, 93. Amphibole: Photo by Patrick Nurre, 93. Amphibolite: http://www.bing.com/images/search?q=Amphibolite+Metamorphic+Rock&FORM=RESTAB&qft=%2bfilterui%3alicense-L2_L3#view=detail&id=5661C7BA00DAEC97138405781B9342A327446253&selectedIndex=1, http://creativecommons.org/licenses/by-sa/3.0/legalcode, CC by-SA 3.0, 94. Minerals: Photos by Heidi Ann Noggle, 94. Rhyolite: Photo by Patrick Nurre, 95. Rhyolite lava: Photo by Heidi Ann Noggle, from author's personal collection, 95. Upper Falls: Public Domain. http://en.wikipedia.org/wiki/Yellowstone_Falls#mediaviewer/File:Upper_Falls_of_the_Yellowstone_River.jpg, 96. Lower Falls: Photo by Vicki Nurre, 96. Canyon Walls: Photo by Vicki Nurre, 96. Potassium Feldspar: Photo by Heidi Ann Noggle, from author's personal collection, 97. Trachyte Lava: Photo by Patrick Nurre, 97. Quartz: Photo by Heidi Ann Noggle, from author's collection, 97. Obsidian: Photo by Vicki Nurre, 98. Obsidian: Photo by Patrick Nurre, 98. Obsidian Cliff: Photo by Patrick Nurre, 99. Obsidian Boulder: Photo by Patrick Nurre, 99. Obsidian Cliff: Photo by Patrick Nurre, 99. Obsidian "Sand": Photo by Patrick Nurre, 100. Vitrophyre: Photo by Patrick Nurre, 100. Vitrophyre: Photo by Patrick Nurre, 100. Pyroclastic Tuff: Photo by Heidi Ann Noggle, from author's personal collection, 101. Pyroclastic Tuff: Heidi Ann Noggle, from author's personal collection, 101. Feldspar Minerals: Photo by Patrick Nurre, 102. Tower Fall: Photo by Vicki Nurre, 102. Quartz and pyroxene: Photo by Heidi Ann Noggle, used by permission, 103. Olivine and Hornblende: Photo by Vicki Nurre, 103. Rhyolite Tuff: Photo by Patrick Nurre, 103. Rhyolite Tuff: Photo by Patrick Nurre, 103. Huckleberry Ridge Tuff: Photo by Patrick Nurre, 104. Mesa Falls Tuff: Public Domain. Found at http://en.wikipedia.org/wiki/Mesa_Falls_Tuff#mediaviewer/File:Ashton_Quarry.jpg, 104. Lava Creek Tuff: Public Domain. Found at http://en.wikipedia.org/wiki/Lava_Creek_Tuff#mediaviewer/File:Tuff_cliff_yellowstone_national_park.jpg, 105. Magnetite: Photo by Vicki Nurre, 105. Olivine: Photo by Vicki Nurre, 105. Calcium Feldspar: Photo by Vicki Nurre, 105. Pyroxene: http://commons.wvc.edu/rdawes/G101OCL/Basics/BscsTables/minerals.html, http://creativecommons.org/licenses/by/3.0/us/legalcode, CC by 3.0, 103. Basalt: Photo by Patrick Nurre, 105. Sheepeater Cliff: Photo by Patrick Nurre, 106. Basalt Tuff: Photo by Patrick Nurre, 106. Breccia and Conglomerate: Photo by Patrick Nurre, 107. Breccia and Conglomerate: Photo by Patrick Nurre, 107. Geyserite: Photo by Vicki Nurre, 108. Mammoth Hot Springs: Photo by Vicki Nurre, 109. Travertine: Photo by Patrick Nurre, 109. Panorama: Photo by Patrick Nurre, 110. Specimen Ridge: Photo courtesy of John Meyer, 110. Petrified Wood: Photo by Heidi Ann Noggle, from author's personal collection, 111. The Great Unconformity: Image by Vicki Nurre, 111. **Rocks of the Greater**

Yellowstone Region from the Author's Collection: Limestone: Photo by Heidi Ann Noggle, from author's personal collection, 112. The Great Unconformity: Photo by Patrick Nurre, 112. Amygdaloidal Basalt: Photo by Heidi Ann Noggle, from author's personal collection, 113. Aphanitic Basalt: Photo by Patrick Nurre, from author's personal collection, 113. Basalt Porphyry: Photo by Heidi Ann Noggle, from author's personal collection, 114. Andesite Porphyry: Photo by Heidi Ann Noggle, from author's personal collection, 114. Diabase: Photo by Heidi Ann Noggle, from author's personal collection, 115. Amphibolite: Photo by Heidi Ann Noggle, from author's personal collection, 115. Hornblende Dacite: Photo by Heidi Ann Noggle, from author's personal collection, 116. Tuff: Photo by Heidi Ann Noggle, from author's personal collection, 116. Tuff: Photo by Heidi Ann Noggle, from author's personal collection, 117. Tuff: Photo by Heidi Ann Noggle, from author's personal collection, 117. Tuff: Photo by Heidi Ann Noggle, from author's personal collection, 118. Tuff: Photo by Heidi Ann Noggle, from author's personal collection, 118. Rhyolite Porphyry: Photo by Heidi Ann Noggle, from author's personal collection, 119. Samples of Rhyolite: Photos by Heidi Ann Noggle, from author's personal collection, 119. Obsidian: Photo by Heidi Ann Noggle, from author's personal collection, 120. Schist: Photo by Heidi Ann Noggle, from author's personal collection, 120. Vesicular Basalt: Photo by Heidi Ann Noggle, from author's personal collection, 121. Dog Tooth" Calcite: Photo by Heidi Ann Noggle, from author's personal collection, 121.

Chapter Eight
Earth: http://en.wikipedia.org/wiki/Ice_age#mediaviewer/File:IceAgeEarth.jpg, http://creativecommons.org/licenses/by/3.0/us/legalcode, CC by 3.0, 123. Fjords: Public Domain. http://en.wikipedia.org/wiki/Ice_age#mediaviewer/File:Scandinavia.TMO2003050.jpg, 124. Volcanic Ash: Photo by Patrick Nurre, 127. Glacial Horn: Photo by Patrick Nurre, 128. Glacial Valley: Photo by Patrick Nurre, 128. Glacial Kettles: Photo by Patrick Nurre, 129. Glacial Horns: Photo by Patrick Nurre, 129. Glacial Erratics: Photo by Patrick Nurre, 130. Glacial Erratics: Photo by Patrick Nurre, 130. Former Lake Yellowstone: Photo by Patrick Nurre, 130. Yellowstone Park Sign: Photo by Patrick Nurre, 131. Water Gap: Photo by Patrick Nurre, 132. Past Water Levels: Photo by Patrick Nurre, 132. Glacial Cirque: Photo by Patrick Nurre, 132. Tumbles and Rounded Cobbles: Photo by Patrick Nurre, 133. Teton Valley: Photo by Patrick Nurre, 133. Glacial Valley: Photo by Patrick Nurre, 133. Glacial Cirques, Tetons: Photo by Patrick Nurre, 134. Glacial Moraine: Photo by Patrick Nurre, 134. Grand Canyon of the Yellowstone: Photo by Vicki Nurre, 135.

Chapter Nine
Map: Found at www.yellowstone-natl-park.com/gethere.htm, 136. Mini photo: Photo by Vicki Nurre, 137. Lewis and Clark Caverns: Photo by Patrick Nurre, 138. Lewis and Clark Caverns: Photo by Patrick Nurre, 138. Folded Limestone: Photo by Vicki Nurre, 139. Dolomite attachment: Photo by Vicki Nurre, 141. Quake Lake: Photo by Brad Dennis, used by permission, 141. Madison River and dam: Photo by Brad Dennis, used by Permission, 142. Collapsed House: Photo by Patrick Nurre, 142. Gallatin Canyon: Photo by Patrick Nurre, 143. Gneiss: Photo by Patrick Nurre, 143. Yellowstone Map: Public Domain. USGS, found at http://www.yellowstone.co/roads.htm, 144. Tuff Wall: Photo by Patrick Nurre, 145. Elk: Photo by Vicki S. Nurre, 145. Buffalo: Photo courtesy of Brad Dennis, 145. Trumpeter Sawn: Public Domain. Photographer unknown. Found at http://www.nps.gov/features/yell/slidefile/birds/waterfowl/Images/00982.jpg, 146. Madison River: http://www.flickr.com/photos/spendadaytouring/4366146518/, https://creativecommons.org/licenses/by/2.0/legalcode, CC by SA 2.0, 146. Firefighter: Photo by Jeff Henry. Courtesy NPS. Found at http://www.nps.gov/features/yell/slidefile/fire/fightingff88/Images/12423.jpg, 147. Map of 1988 fires: Courtesy NPS. http://www.nps.gov/features/yell/slidefile/fire/graphics88/Images/13456.jpg, 147. Burned area: Photo by Jim Peaco. Courtesy NPS. Found at http://www.nps.gov/features/yell/slidefile/fire/postfiresuccession88/Images/13088.jpg, 148. New growth: Photo by Brian Suderman. Courtesy of NPS. Found at http://www.nps.gov/features/yell/slidefile/fire/postfiresuccession88/Images/16160.jpg, 148. New growth: Photo courtesy Brad Dennis, 149. Mini photo: Photo by Vicki Nurre, 149.Grizzly Bear marks: Photo by Vicki S. Nurre, 150. Gibbon Falls: Photo by Vicki S. Nurre, 150. Beryl Spring: Photo by Thomas Duesing, http://www.flickr.com/photos/tfduesing/195711338/, https://creativecommons.org/licenses/by/2.0/legalcode, CC by 2.0, 151. Monument Geyser Basin: Public Domain. Found at http://mms.nps.gov/yell/ofvec/exhibits/treasures/monument/index.htm, 151. Artist Paint Pots and "Bobby-socks": Photo by Patrick Nurre, 152. Map of Norris Geyser Basin: http://yellowstone.net/geysers/files/2011/02/norrismap_2.gif, 153. Back Basin: Photo by Patrick Nurre, 154. Porcelain Basin: Photo by Patrick Nurre, 154. Echinus Geyser: Public Domain. Found at http://en.wikipedia.org/wiki/Echinus_Geyser#mediaviewer/File:Echinus_geyser.jpg, 155. Steamboat Geyser: http://en.wikipedia.org/wiki/Steamboat_Geyser#mediaviewer/File:Steamboat_Geyser.jpg, 155. Steamboat Geyser: http://en.wikipedia.org/wiki/Steamboat_Geyser#mediaviewer/File:NorrisGeyserBasinSteamboat.JPG, 155. Steamboat geyser: Photo by Vicki Nurre, 156. Yellowstone Map: Public Domain. USGS, found at http://www.yellowstone.co/roads.htm, 157. Roaring Mountain 1942: Public Domain. Found at http://en.wikipedia.org/wiki/Roaring_Mountain#mediaviewer/File:Yellowstone-Roaring_Mountain_Aat10.jpg, 158. Roaring Mountain 2010: http://en.wikipedia.org/wiki/Roaring_Mountain#mediaviewer/File:Roaring_Mountain_YNP1.jpg, http://creativecommons.org/licenses/by/3.0/us/legalcode, CC by 3.0, 158. Obsidian Cliff: Photo by Patrick Nurre, 159. Obsidian: Photo by Vicki Nurre, 159. Sheepeater Cliff: Photo courtesy of Brad Dennis, 160. Basalt Lava Flow: Photo by Vicki Nurre, 161. Bunsen Peak: Public Domain. http://en.wikipedia.org/wiki/Bunsen_Peak#mediaviewer/File:BunsenPeakFromSwanLake1922.jpg, 162. Hypothetical Diagram of Lacolith: Public Domain. http://commons.wikimedia.org/wiki/File:Laccolith.svg, 162. Huckleberry Ridge Tuff: https://c2.staticflickr.com/8/7264/7654402758_f1dc1af75b_z.jpg, 163. Golden Gate Bridge after earthquake: Courtesy NPS, found at http://www.nps.gov/features/yell/slidefile/geology/1959earthquake/Images/02683.jpg, 163. Mammoth Terraces: Photo by Vicki Nurre, 165. Active Hot Springs, Mammoth Terrace: Photo by Vicki Nurre, 165. Liberty Cap: Photo courtesy Brad Dennis, 166. Active hot springs: Photo by Vicki Nurre, 166. Mountain Goat: Photo by Patrick Nurre, 167. Roosevelt Arch: http://en.wikipedia.org/wiki/Roosevelt_Arch#mediaviewer/File:Yellowstone_North_Gate.jpg, http://creativecommons.org/licenses/by/3.0/us/legalcode, CC by 3.0, 167. Teddy Roosevelt: Public Domain. Found at

http://en.wikipedia.org/wiki/Theodore_Roosevelt#mediaviewer/File:Theodore_Roosevelt_by_John_Singer_Sargent,_1903.jpg, 168. Yellowstone River: Public Domain. Found at http://en.wikipedia.org/wiki/Paradise_Valley_(Montana)#mediaviewer/File:Yellowstone_River,_flowing_through_Paradise_Valley.jpg, 169. Electric Peak: Photo by Patrick Nurre, 169. Drumlin Field: Public Domain. Found at http://en.wikipedia.org/wiki/Drumlin#mediaviewer/File:Drumlinfield_large.jpg, 170. Glacial Morraine: Public Domain. Found at http://www.nps.gov/features/yell/slidefile/geology/glacial/index.htm, 170. Glacial Till: Photo by Patrick Nurre, 171. Specimen Ridge 1890: http://en.wikipedia.org/wiki/Specimen_Ridge#mediaviewer/File:FossilTreesSpecimenRidgeYNP1890.jpg, 172. Specimen Ridge 2013: Photo courtesy of John Meyer, 172. Mini photo: Photo courtesy of John Meyer, 172. Fossil Wood Stump: Photo courtesy of John Meyer, 173. Fossil wood stump: Photo courtesy of John Meyer, 173. Fossil wood stump: Photo courtesy of John Meyer, 173. Tuff Boulder: Photo by Patrick Nurre, 174. The Great Unconformity: Photo by Patrick Nurre, 175. Glacial Erratics: Photo by Patrick Nurre, 176. Map Beartooth Highway: Public Domain. Found at http://en.wikipedia.org/wiki/Beartooth_Highway#mediaviewer/File:Beartooth_Highway_map.svg, 176. Cooke City: Photo by Ildar Sagdejev, free to use, http://en.wikipedia.org/wiki/Cooke_City-Silver_Gate,_Montana#mediaviewer/File:2003-08-17_Miners_Saloon_in_Cooke_City,_Montana.jpg, 178. Snowpack: Photo by Vicki Nurre, 179. Switchbacks: Photo by Vicki Nurre, 179. Mini photos: Photos by Vicki Nurre, 179. Glacial Valley, Beartooth Mountains: Patrick Nurre, 180. Uplifted Mountain: Photo by Patrick Nurre, 181. Heart Mountain: http://en.wikipedia.org/wiki/Heart_Mountain_(Wyoming)#mediaviewer/File:Heart_Mountain_Wyoming.jpg http://creativecommons.org/licenses/by-sa/2.0/legalcode, CC by-SA 2.0, 182. Tower Fall: Photo by Vicki Nurre, 183. Unconformities at Tower Fall: Photo by Vicki Nurre, 183. Yellowstone Map: Public Domain. USGS, found at http://www.yellowstone.co/roads.htm, 184. Grand Canyon of the Yellowstone: Photo by Vicki S. Nurre, 185. Glacial Lake Yellowstone: apushnationalzoo.wikispaces.com, 186. Sulpher Cauldron with Sour Creek Dome: Photo by Patrick Nurre, 186. Mud Volcano: Photo by Patrick Nurre, 187. Black Dragon's Caldron: Photo courtesy of Brad Dennis, 187. Fishing Bridge 1950's: http://www.bing.com/images/search?&q=fishing+bridge&qft=+filterui:license-L2_L3&FORM=R5IR42#view=detail&id=6C8F3DC651269083AD37B2317FD61457DEF259EA&selectedIndex=2, 188. Yellowstone Map: Public Domain. USGS, found at http://www.yellowstone.co/roads.htm, 189. Yellowstone Lake: Photo by Vicki Nurre, 190. West Thumb Geyser Basin: Photo by Vicki Nurre, 190. Fishing Cone 1928: http://en.wikipedia.org/wiki/Fishing_Cone#mediaviewer/File:FishingConePostcard-Asahel_Curtis1928.jpg, 191. Fishing Cone: http://en.wikipedia.org/wiki/Fishing_Cone#mediaviewer/File:Fishing_cone.jpg, 192. Fishing Cone Under Water: http://en.wikipedia.org/wiki/Fishing_Cone#mediaviewer/File:Fishing_Cone_071610.jpg, 192. West Thumb Geyser Basin thermophiles: Photo by Vicki Nurre, 193. West Thumb Geyser Basin Thermophiles: Photo by Patrick Nurre, 193. Isa Lake: http://en.wikipedia.org/wiki/Isa_Lake#mediaviewer/File:Isa_Lake_YNP1.jpg, http://creativecommons.org/licenses/by-sa/3.0/legalcode, CC by-SA 3.0, 194. Lily pad: Public domain. Found at http://www.nps.gov/features/yell/slidefile/plants/waterlily/Images/07632.jpg, 194. Lone Star Geyser: Photo by Mark Wagner, http://en.wikipedia.org/wiki/Lone_Star_Geyser#mediaviewer/File:Lone_Star_Geyser_pre-eruptive_phase_20080812.jpg, http://creativecommons.org/licenses/by/2.5/legalcode, CC by-2.5, 195. Yellowstone Map: Public Domain. USGS, found at http://www.yellowstone.co/roads.htm, 196. Old Faithful Map: Public Domain. Found at http://en.wikipedia.org/wiki/Old_Faithful#mediaviewer/File:SouthernSectionUpperGeyserBasinOldFaithful.JPG, 197. Mini photo: Photo courtesy of Brad Dennis, 197. Old Faithful and Geyser Hill: Photo by Patrick Nurre, 198. Sapphire Pool: Photo by Frank Kovalchek, http://www.flickr.com/photos/72213316@N00/3652341573/, https://creativecommons.org/licenses/by/2.0/legalcode, CC by 2.0, 199. Sawmill and Grand Geysers: http://en.wikipedia.org/wiki/Geothermal_areas_of_Yellowstone#mediaviewer/File:Sawmill_and_Grand_geysers_erupting_in_Yellowstone_NP.jpg, http://creativecommons.org/licenses/by/3.0/us/legalcode, CC by 3.0, 200. Grand Geyser: http://en.wikipedia.org/wiki/Grand_Geyser#mediaviewer/File:GrandGeyserYNP.jpg, https://creativecommons.org/licenses/by/2.0/legalcode, CC by 2.0, 200. Old Faithful: Public Domain. Found at http://en.wikipedia.org/wiki/Old_Faithful#mediaviewer/File:Old_Faithful_geyser,_Yellowstone_National_Park_-_NARA_-_517017.jpg, 201. Old Faithful: Photo by Jim Peaco. Public Domain. Found at http://www.nps.gov/features/yell/slidefile/thermalfeatures/geysers/upper/Images/17530.jpg, 201. Daisy Geyser: Photo by Brocken Inaglory. http://en.wikipedia.org/wiki/Daisy_Geyser#mediaviewer/File:Daisy_Geyser_erupting_in_Yellowstone_National_Park_edit.jpg, http://creativecommons.org/licenses/by/3.0/us/legalcode, CC by 3.0, 201. Riverside Geyser: Photo by Vicki S. Nurre, 202. Morning Glory Pool: Public Domain. Found at http://en.wikipedia.org/wiki/Morning_Glory_Pool#mediaviewer/File:Morning_Glory_Pool.jpg, 203. Mystic Falls: http://en.wikipedia.org/wiki/Mystic_Falls#mediaviewer/File:MysticFallsYNP.jpg, https://creativecommons.org/licenses/by/2.0/legalcode, CC by 2.0, 203. Acidothermophiles: Photo by Vicki Nurre, 204. Run-off at Firehole River: Photo by Vicki Nurre, 204. Great Fountain Geyser: Public Domain. Found at http://en.wikipedia.org/wiki/Great_Fountain_Geyser#mediaviewer/File:GreatFountainGeyser-Wegner1967.jpg, 205. Firehole Lake Drive: Public Domain. Found at http://www.nationalparkstraveler.com/files/yell-firehole_lake_drive_rd_damage_nps_600_7-7-2014.jpg, 205. Fountain Paint Pots map: Public Domain. Found at http://en.wikipedia.org/wiki/Fountain_Paint_Pot#mediaviewer/File:FountainPaintPotsMap-Fountain_Paint_Pot.JPG, 206. Fountain paint Pot: Public Domain. Found at http://en.wikipedia.org/wiki/Fountain_Paint_Pot#mediaviewer/File:Fountain_paint_pots_yellowstone.jpg, 208. Red Spouter: Courtesy NPS. Found at http://www.nps.gov/features/yell/slidefile/thermalfeatures/mudpots/midwaylower/Images/05406.jpg, 208. Silex Spring: http://en.wikipedia.org/wiki/Geothermal_areas_of_Yellowstone#mediaviewer/File:Silex_spring_in_yellowstone.jpg, http://creativecommons.org/licenses/by/3.0/us/legalcode, CC by 3.0, 208. Silex Spring: Photo by Brocken Inaglory. http://en.wikipedia.org/wiki/Geothermal_areas_of_Yellowstone#mediaviewer/File:Silex_spring_overflow_in_yellowstone.jpg, http://creativecommons.org/licenses/by/3.0/us/legalcode, CC by 3.0, 208. Fairy Falls: http://www.flickr.com/photos/soupaloop/11110982126/, https://creativecommons.org/licenses/by/2.0/legalcode, CC by 2.0, 209.

Patrick Nurre has been a rock hound since childhood and has an extensive rock, mineral and fossil collection, having collected from all over the United States. In 2005, he started Northwest Treasures, which is devoted to designing geology kits for schools. He conducts numerous field trips each year in Washington State to such places as the Olympic Peninsula, Mt. Rainier, Mt. St. Helens, the Channeled Scablands, Mt. Baker and Whidbey Island. In addition, he also gives field trips to the volcano loop of Oregon and California, Mt. Hood volcanic area (Oregon), the eastern badlands of Montana and Yellowstone National Park. In 2012, he opened the Geology Learning Center in Mountlake Terrace, WA. He is a popular speaker at homeschool conventions, schools, and churches. Patrick currently co-pastors a church in the Seattle, Washington area.

If you would like to contact Patrick about speaking or field trips: northwestexpedition@msn.com For a list of speaking topics: NorthwestRockAndFossil.com

Other books by Patrick Nurre – these are all also available with sample rock, mineral, and fossil kits at NorthwestRockAndFossil.com.

Rocks and Minerals for Little Eyes (PreK-3)
Fossils and Dinosaurs for Little Eyes (PreK-3)
Volcanoes for Little Eyes (PreK-3)
Geology for Kids (3-6)
Rock Identification Made Easy (3-12)
Fossil Identification Made Easy (3-12)
Bedrock Geology (high school)
Rocks and Minerals: The Stuff of the Earth (high school)
Volcanoes, Volcanic Rocks and Earthquakes (high school)
The Geology of Yellowstone – A Biblical Guide
Genesis Rock Solid – A Biblical View of Geology
Fossils, Dinosaurs and Cave Men

CPSIA information can be obtained
at www.ICGtesting.com
Printed in the USA
FSOW03n1716020117
29089FS